# Success
## Assessment Papers

# Maths

### 7 – 8 years

Paul Broadbent

# Sample page

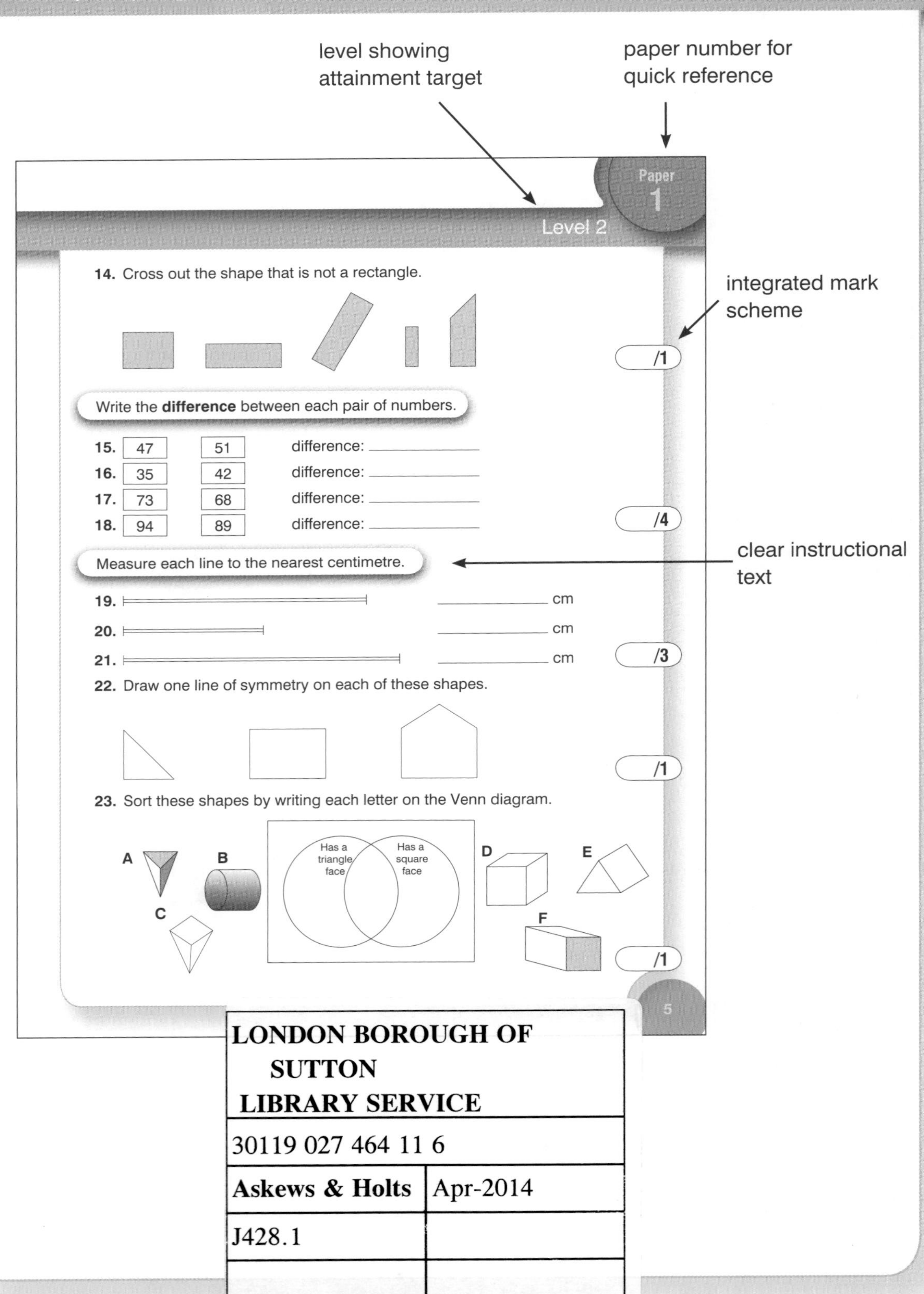
level showing attainment target
paper number for quick reference
integrated mark scheme
clear instructional text
Paper 1
Level 2
14. Cross out the shape that is not a rectangle.
/1
Write the difference between each pair of numbers.
15. 47 51 difference:
16. 35 42 difference:
17. 73 68 difference:
18. 94 89 difference:
/4
Measure each line to the nearest centimetre.
19. cm
20. cm
21. cm
/3
22. Draw one line of symmetry on each of these shapes.
/1
23. Sort these shapes by writing each letter on the Venn diagram.
A
B
C
D
E
F
Has a triangle face
Has a square face
/1
5

# Contents

## PAPER 1

**1.** Circle the missing number.

83 = ____ + 3

**a)** 8 **b)** 80 **c)** 11 **d)** 5

/1

Write the next two numbers in each **sequence.**

**2.** 12 14 16 18 20 ____ ____

**3.** 35 33 31 29 27 ____ ____

**4.** 41 43 45 47 49 ____ ____

**5.** 64 62 60 58 56 ____ ____

/4

**6.** Tick the clock that shows 7.15.

**a)** 

**b)** 

**c)** 

**d)** 

/1

Write the fraction shaded on each of these shapes.

**7.** 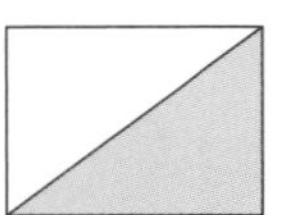

**8.** 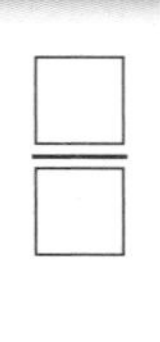

**9.** 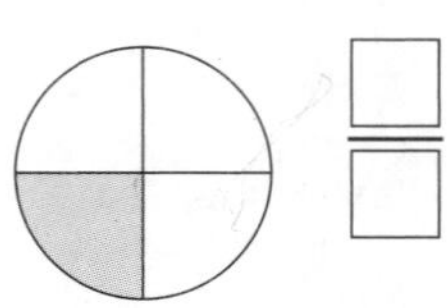

/3

Answer these.

**10.** 12 + 7 = ☐

**11.** 15 + 3 = ☐

**12.** 13 + 6 = ☐

**13.** 11 + 5 = ☐

/4

**14.** Cross out the shape that is not a rectangle.

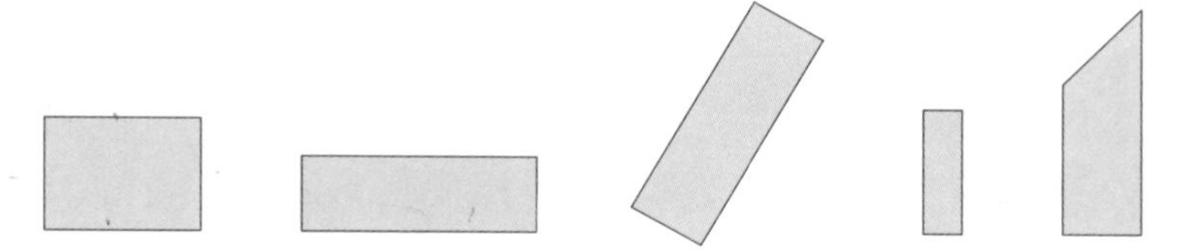

/1

Write the **difference** between each pair of numbers.

| | | | |
|---|---|---|---|
| **15.** | 47 | 51 | difference: ______ |
| **16.** | 35 | 42 | difference: ______ |
| **17.** | 73 | 68 | difference: ______ |
| **18.** | 94 | 89 | difference: ______ |

/4

Measure each line to the nearest centimetre.

**19.** ______ cm

**20.** ______ cm

**21.** ______ cm

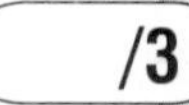

/3

**22.** Draw one line of symmetry on each of these shapes.

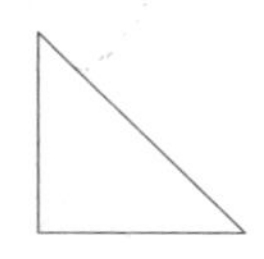

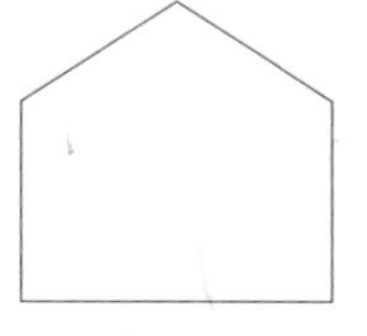

/1

**23.** Sort these shapes by writing each letter on the Venn diagram.

A

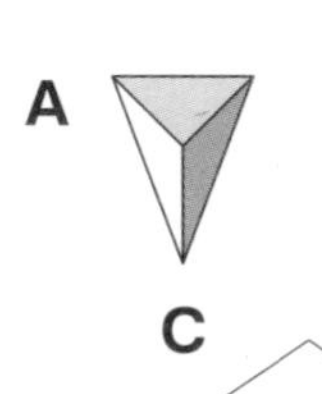

B

C

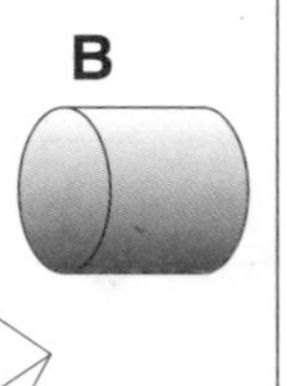

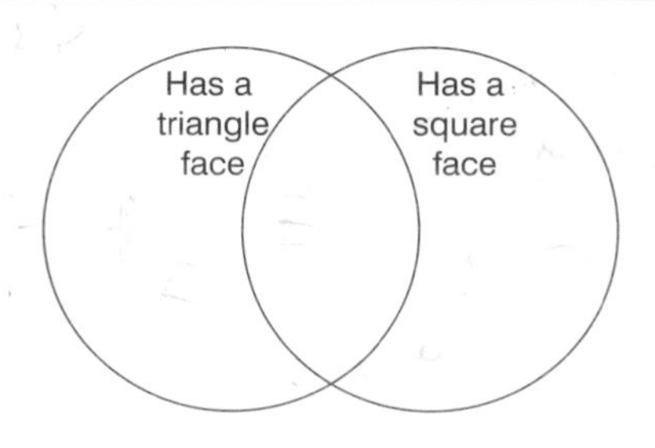


D

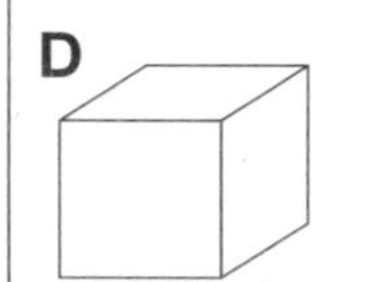

E

F

/1

Answer these.

**24.** $6 \times 3 =$ ______

**25.** $7 \times 2 =$ ______

/2

**26.** Circle the missing number.

______ $\div 2 = 4$

**a)** 8 **b)** 2 **c)** 6 **d)** 12

/1

Write the missing numbers or words.

**27.** 91 → ______________________

**28.** ______ → fifty-six

**29.** 83 → ______________________

**30.** ______ → seventy-four

/4

/30

## PAPER 2

Answer these.

**1.** Which day comes before Thursday? ______________

**2.** Which day comes after Friday? ______________

**3.** Which day is two days after Sunday? ______________

**4.** Which day is two days before Wednesday? ______________

/4

Circle the correct shapes in each row.

**5.** square

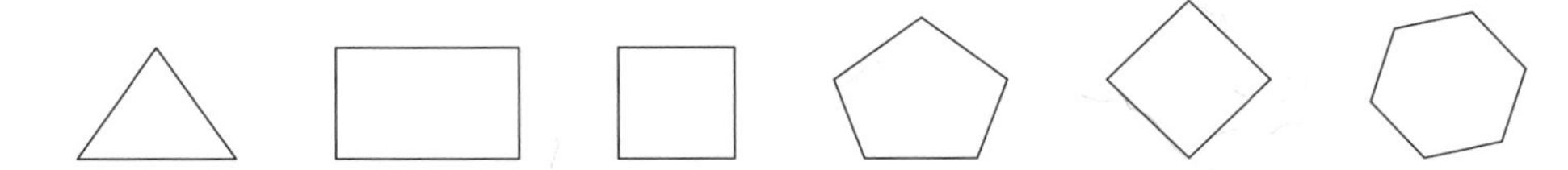

**6.** triangle

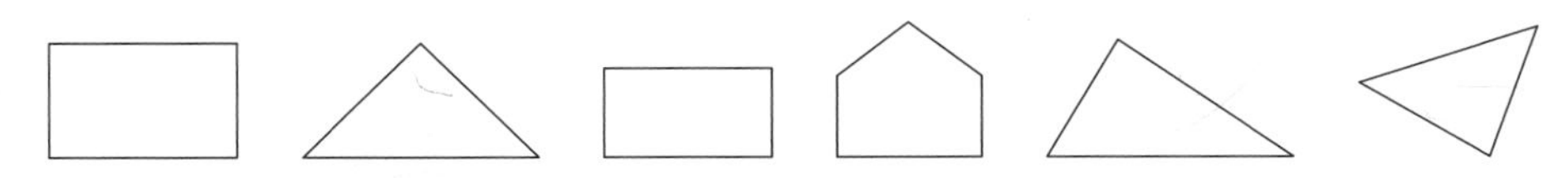

**7.** hexagon

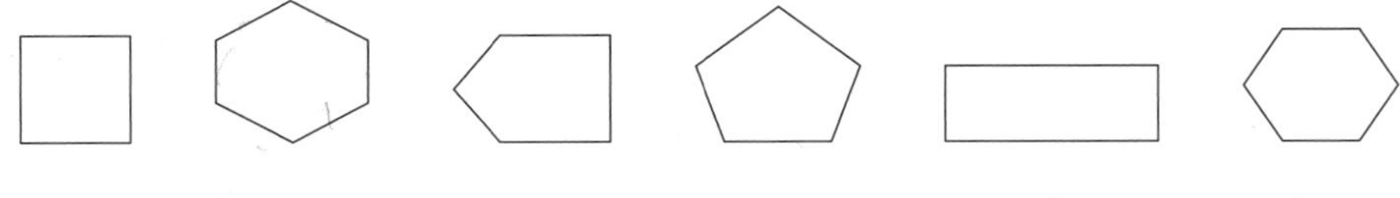

**8.** pentagon

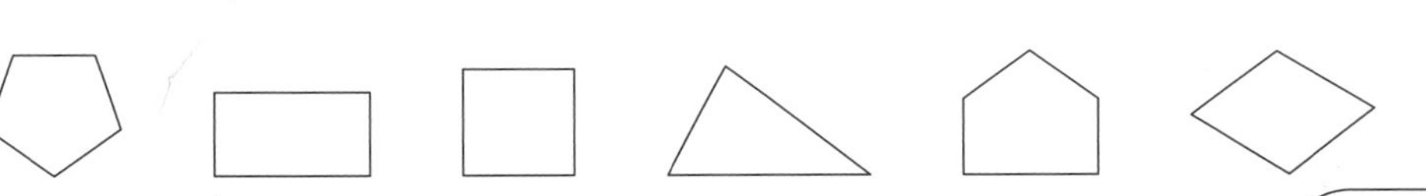

/4

**9.** Write the total.

£

/1

**10.** What is the **difference** between these two weights?

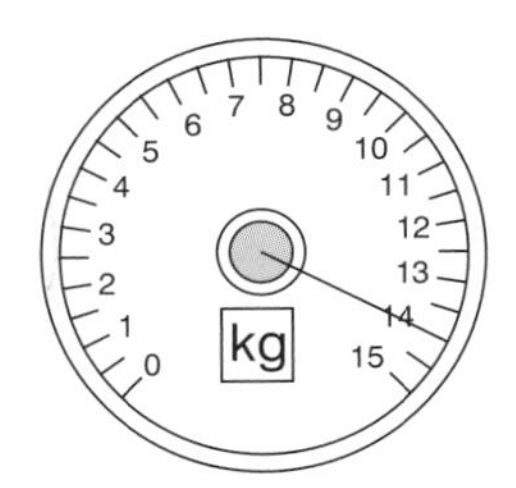


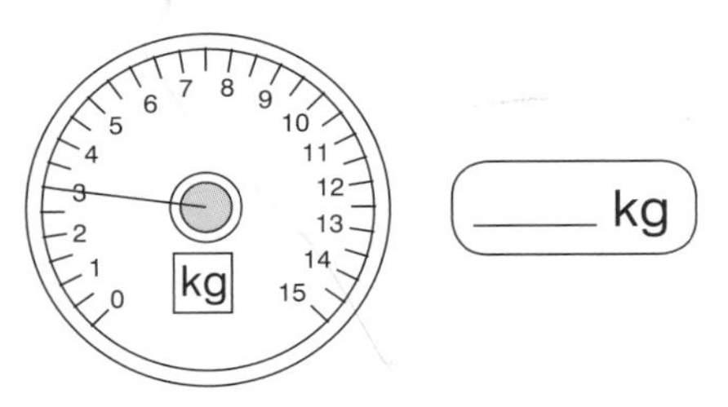


_____ kg

/1

Answer these.

**11.** 2 × 6 = _____

**12.** 3 × 5 = _____

**13.** 8 × 10 = _____

**14.** 9 × 2 = _____

/4

**15.** Finish drawing the jumps on the number line and write the missing numbers.

/1

Answer these.

**16.** What is 17 subtract 9? ______

**17.** What is the total of 8 and 7? ______

**18.** What is 19 subtract 13? ______

**19.** What is 6 more than 15? ______

/4

Colour half of each set of beads.
Write the answer.

**20.**

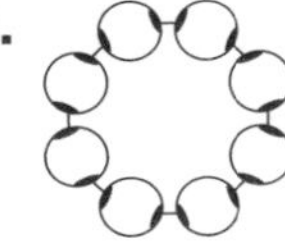

$\frac{1}{2}$ of 8 = ______

**21.**

$\frac{1}{2}$ of 12 = ______

**22.** Write the number 87 in words. ______

/3

Name these shapes.

**23.**

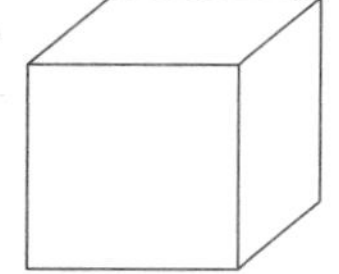

______

**24.**

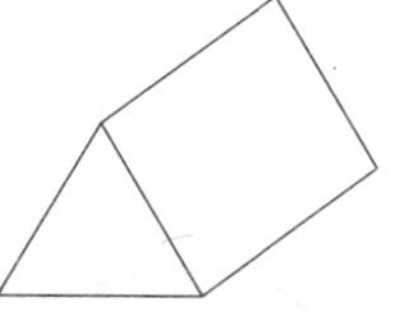

______

**25.**

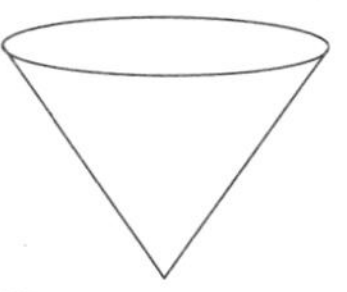

______

/3

Answer these.

**26.** 30 + 50 = ______

**27.** 90 − 70 = ______

**28.** 70 − 30 = ______

**29.** 80 + 80 = ______

**30.** 60 + 90 = ______

/5

/30

## PAPER 3

Answer these.

**1.** 47 + 30 = ______

**2.** 79 + 20 = ______

**3.** 54 + 40 = ______

**4.** 35 + 50 = ______

**5.** Tick the triangle that is **symmetrical**.

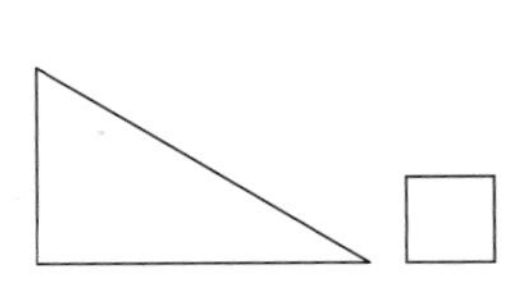 ☐ 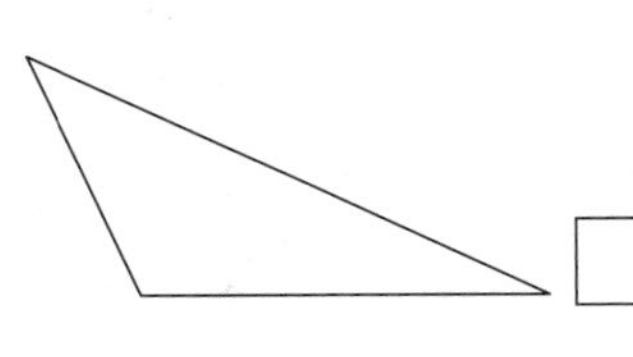 ☐ 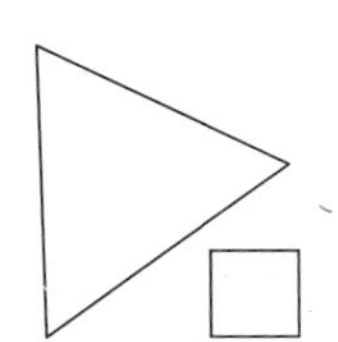 ☐ 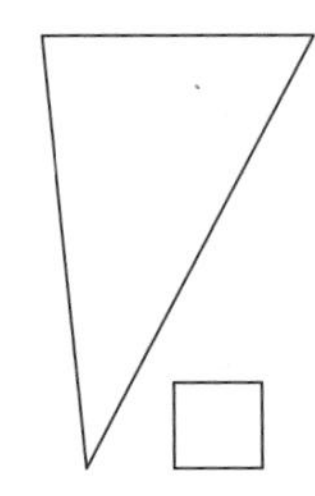 ☐

/1

Write the missing numbers.

**6.** ☐ × 4 = 20

**7.** 15 ÷ ☐ = 5

**8.** 9 × ☐ = 18

**9.** ☐ ÷ 3 = 7

/4

**10.** Colour $\frac{1}{3}$ of this grid.

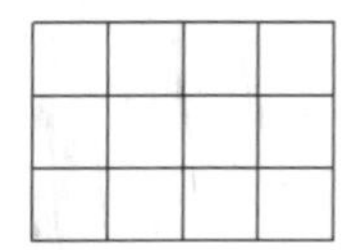

/1

**11.** Finish drawing the jumps on the number line and write the missing numbers.

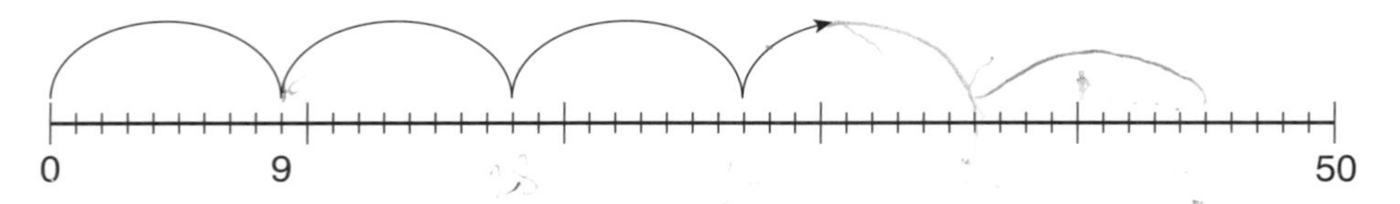


/1

**12.** Into which box will an orange be sorted?

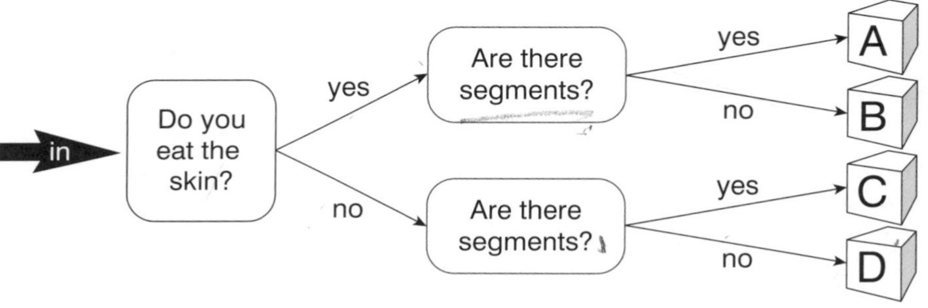


Box ______

/1

**13.** Two eggs cost 20p. What will 6 eggs cost?

______

/1

**14.** Circle the odd numbers in this set.

34 55 27 70 91 85 56

/1

Write these numbers.

**15.** seventy-four → ______

**16.** ninety-one → ______

**17.** eighty-six → ______

/3

This graph shows the bedtimes of a group of children.

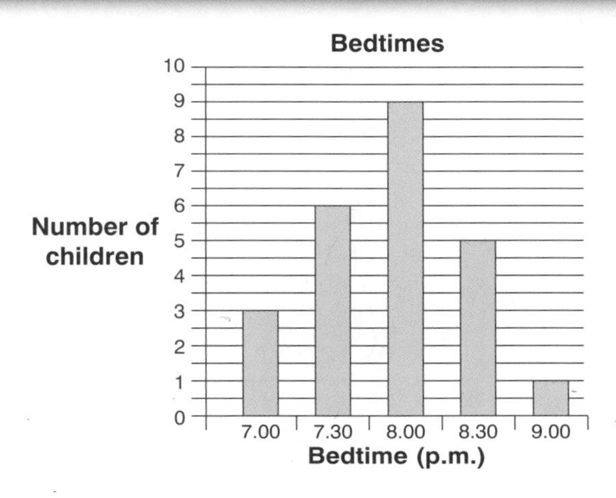


**18.** How many children went to bed at half past seven? ______________

**19.** At which time did three children go to bed? ______________

**20.** How many children altogether went to bed after 8.15? ______________

**21.** How many children altogether went to bed before 7.45? ______________

/4

Measure the length of each line.

**22.** |———————————————| ______________ cm

**23.** |————————————————————| ______________ cm

**24.** What is the difference between the two lengths? _______ cm

**25.** What is 46 rounded to the nearest ten? _______

**26.** What is 84 rounded to the nearest ten? _______

/5

Write the next two numbers in each **sequence**.

**27.** 12 17 22 27 32 _______ _______

**28.** 15 17 19 21 23 _______ _______

**29.** 21 24 27 30 33 _______ _______

**30.** 14 18 22 26 30 _______ _______

/4

/30

# PAPER 4

How much money is in each purse?

1. 

£ ______

2. 

£ ______

3. 

£ ______

4. 

£ ______

/4

Answer these.

5. 18 + 13 = ______

6. 17 + 19 = ______

7. 15 + 11 = ______

8. 18 + 17 = ______

/4

Draw a line of symmetry on each shape.

9. 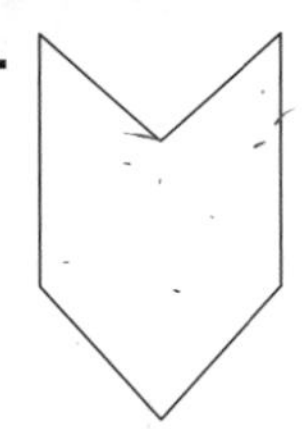

10. 

11. 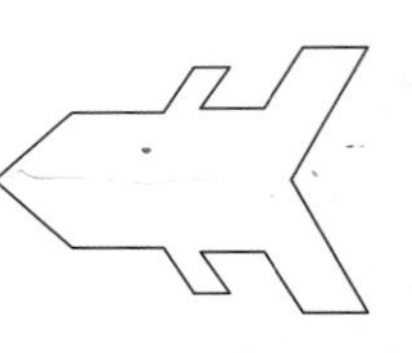

/3

12. Complete the table to show the missing numbers for this machine.

| IN | 15 | | 20 | | 18 |
|---|---|---|---|---|---|
| OUT | | 4 | | 7 | |

/1

Write the times shown on these clocks.

**13.** 

**14.** 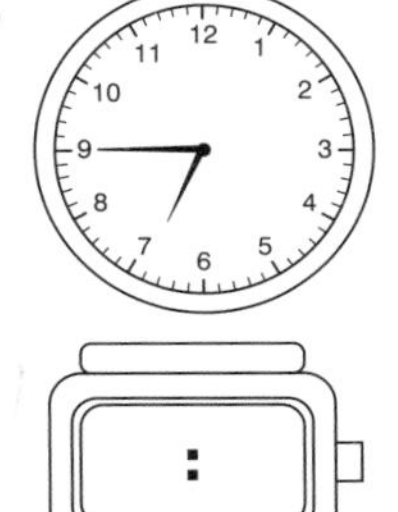

**15.** 

**16.** 

/4

**17.** A bucket holds 6 litres of water. It is poured out equally into 3 jugs. How much water is in each jug?

__________

/1

Write each group of numbers in order starting, with the smallest.

**18.** 19 11 22 15 __________

**19.** 23 17 18 21 __________

**20.** What is $\frac{1}{4}$ of £12? __________

**21.** What is $\frac{1}{2}$ of £5? __________

/4

Name these shapes.

**22.** 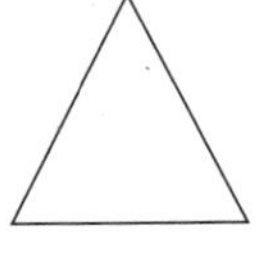

**23.** 

**24.** 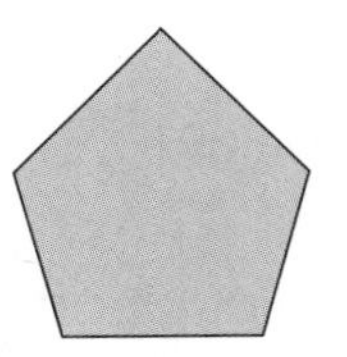

**25.** 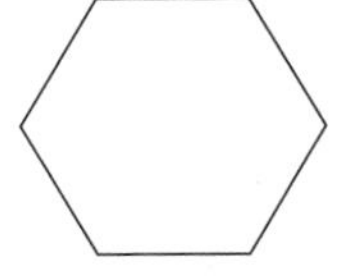

__________ __________ __________ __________

/4

**26.** Circle the even numbers in this set.

35 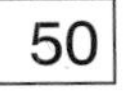 50 67

 92 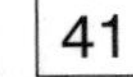 41  28 76

/1

Answer these.

**27.** $2 \times 9 =$ ______ **28.** $3 \times 4 =$ ______

**29.** $5 \times 10 =$ ______ **30.** $6 \times 3 =$ ______ /4

/30

## PAPER 5

**1.** What is the total **area** of this shape? Circle the answer.

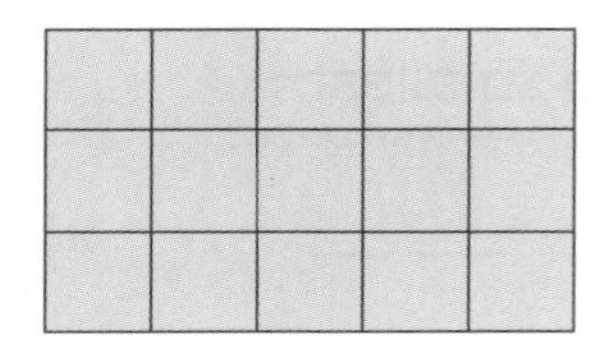

**a)** 5 squares **b)** 18 squares **c)** 15 squares **d)** 10 squares /1

How many hundreds, tens and ones are there are in each of these numbers?

**2.** $165 = 100 +$ ______ $+$ ______

**3.** $219 =$ ______ $+ 10 +$ ______

**4.** $456 =$ ______ $+$ ______ $+ 6$ /3

Write <, > or = for each calculation.

**5.** $(3 + 9)$ ______ $12$ **6.** $(11 - 6)$ ______ $4$

**7.** $(5 + 6)$ ______ $13$ **8.** $(10 - 4)$ ______ $11$

**9.** $(9 + 6)$ ______ $15$ **10.** $(13 - 4)$ ______ $6$ /6

**11.** Draw lines to match the pairs.

3 × 4 9 × 2 1 × 6

6 × 3 2 × 6 3 × 2

/1

Answer these.

**12.** $\frac{1}{2}$ of 14 = ____________

**13.** $\frac{1}{4}$ of 12 = ____________

**14.** $\frac{1}{2}$ of 10 = ____________

/3

Write the missing numbers in each **sequence**.

**15.** 17 ______ 37 ______ 57 67

**16.** 11 14 ______ 20 23 ______

/2

These are the favourite sports of a group of children.

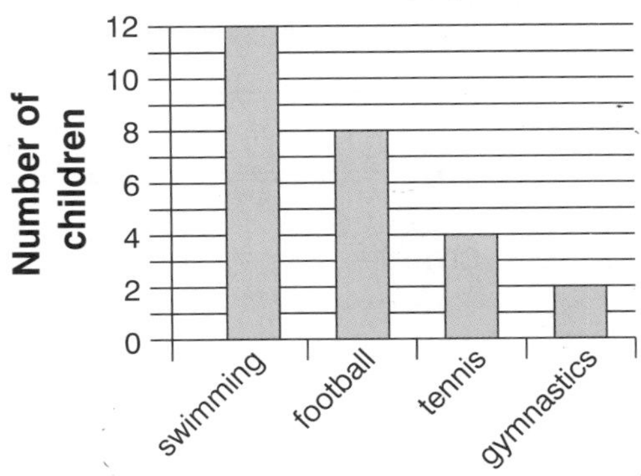


**17.** How many children chose football? ____________

**18.** Which sport was chosen by 4 children? ____________

**19.** How many more children chose swimming than gymnastics? ______________

/3

**20.** Complete the table to show the missing numbers for this machine.

| IN | 18 |  | 25 |  | 23 |
|---|---|---|---|---|---|
| OUT |  | 9 |  | 5 |  |

/1

Write these numbers.

**21.** one hundred and five ______________

**22.** one hundred and fifty ______________

**23.** one hundred and fifteen ______________

/3

Name these shapes.

**24.** 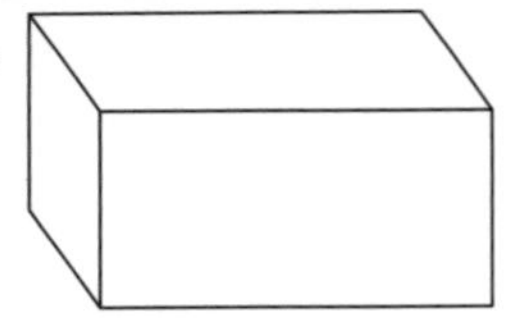

______________

**25.** 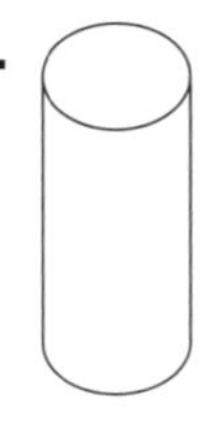

______________

/2

Use a ruler to measure the exact length of each line.
Write each length in centimetres.

**26.** length: ______________ cm

**27.** length: ______________ cm

**28.** length: ______________ cm

/3

Complete these.

/2

**29.** 15 ÷ 5 = ______________

**30.** 18 ÷ 3 = ______________

/30

## PAPER 6

The centre number is the total of the numbers around the outside. Write each centre number.

**1.**

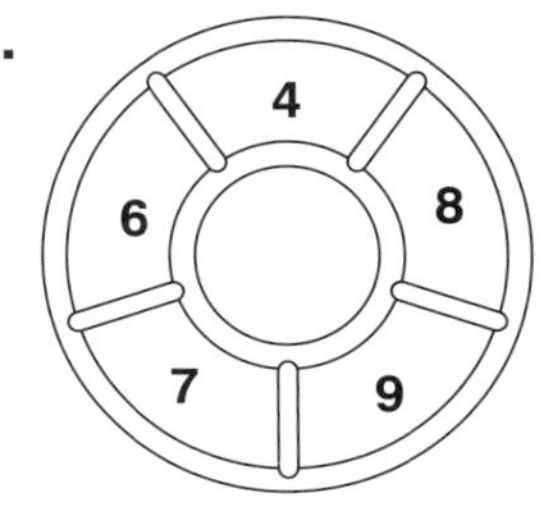


**2.**

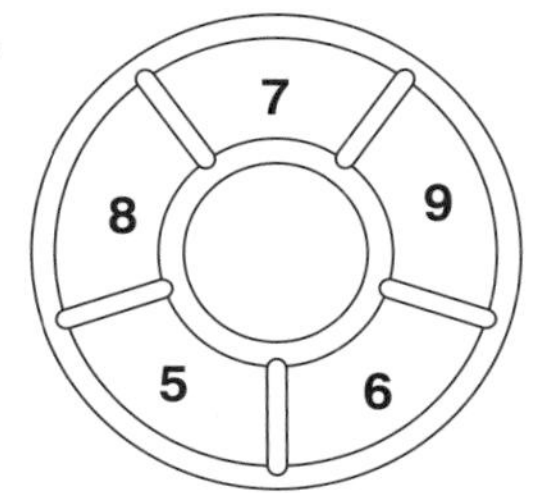


/2

Draw a line of symmetry on each shape.

**3.**

**4.**

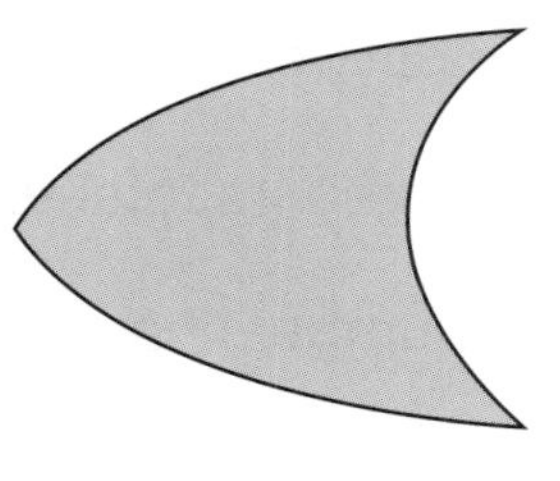

/2

Answer these.

**5.** 45 + 60 = ____________

**6.** 38 + 90 = ____________

**7.** 54 + 70 = ____________

**8.** 93 + 40 = ____________

**9.** 62 + 80 = ____________

/5

## Level 2/3

**Use a ruler to measure the exact length of each line. Write each length in centimetres.**

**10.** length: ________ cm

**11.** length: ________ cm

**12.** length: ________ cm

/3

**Use these numbers to answer the questions.**

35 12 60 84 23

**13.** What is the total of the odd numbers? ________

**14.** What is the difference between the greatest and smallest numbers? ________

**15.** Which two numbers total 95? ________

**16.** Which two numbers have a difference of 11? ________

/4

**Write the number of minutes between each of these times.**

**17.**

________ minutes

**18.**

________ minutes

**19.**

________ minutes

/3

Answer these.

20. How many 100 g weights equal 1 kg? ____________

21. How many 500 g weights equal 1 kg? ____________

22. How many 200 g weights equal 1 kg? ____________

23. How many 50 g weights equal 1 kg? ____________

/4

How much money is in each purse?

24.

£

25.

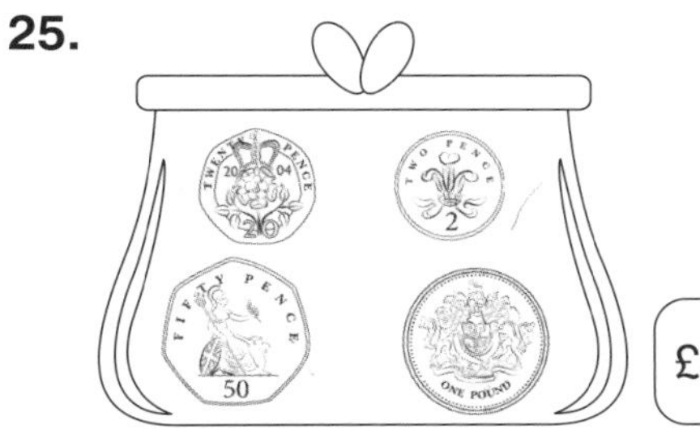

£

/2

Answer these.

26. $\frac{1}{2}$ of 18 = ____________

27. $\frac{1}{2}$ of 22 = ____________

28. $\frac{1}{4}$ of 16 = ____________

29. $\frac{1}{4}$ of 20 = ____________

30. $\frac{1}{3}$ of 12 = ____________

/5

/30

# PAPER 7

Write how many hundreds, tens and ones there are in each of these numbers.

1. 215 = 200 + ______ + ______
2. 193 = ______ + 90 + ______
3. 262 = ______ + ______ + 2
4. 517 = ______ + ______ + ______
5. 321 = ______ + ______ + ______
6. 486 = ______ + ______ + ______

/6

Write the name of each shape.

7. 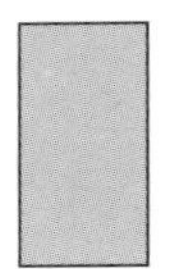

8. 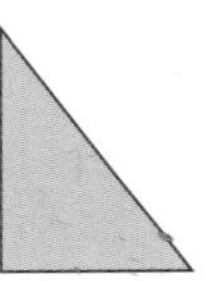

9. 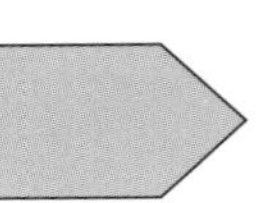

10. 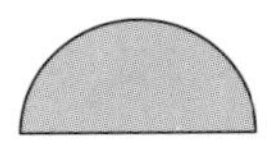

______ ______ ______ ______

/4

What is the change from £1 for each of these prices?

11. 75p change = ____p

12. 89p change = ____p

13. 61p change = ____p

14. 83p change = ____p

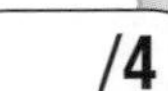

Answer these.

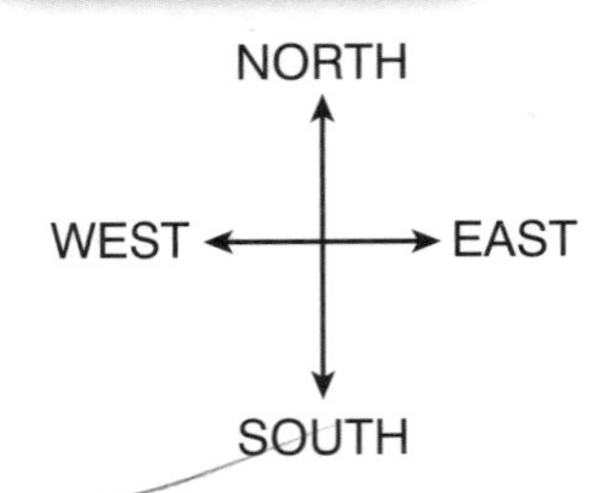


15. Face North and make a quarter turn **clockwise**. In which direction are you facing?

______

**16.** Face South and make a half turn **clockwise**. In which direction are you facing?

__________________

**17.** Face East and make a quarter turn **anticlockwise**. In which direction are you facing?

__________________

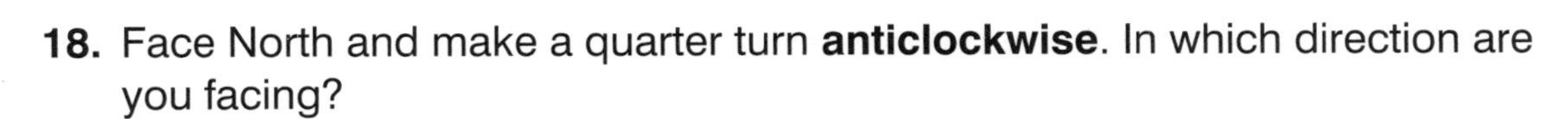

**18.** Face North and make a quarter turn **anticlockwise**. In which direction are you facing?

__________________

/4

**19.** Write the missing number.

_____ $\div 3 = 5$

/1

Draw the reflection of these shapes.

**20.**

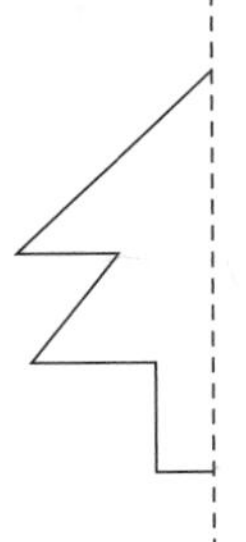

**21.**

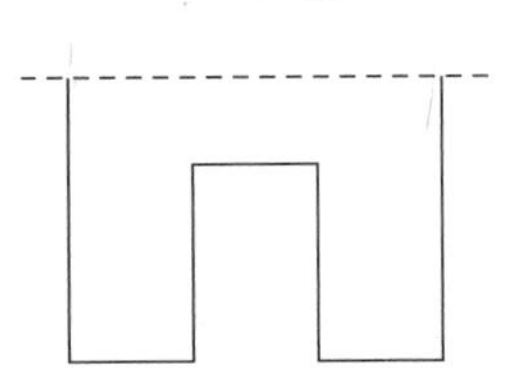

/2

Write these in centimetres.

**22.** 8 metres = __________________ cm

**23.** $6\frac{1}{2}$ metres = __________________ cm

**24.** What is the next even number after 398? __________________

**25.** What is the next odd number after 699? __________________

/4

**26.** Tick the clock that shows 1.45.

a) 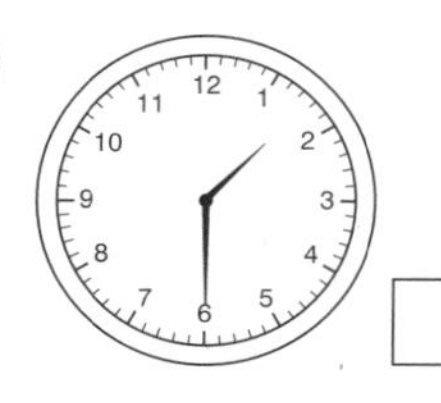 ☐

b) 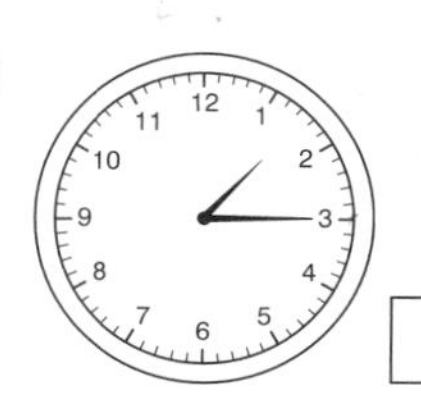 ☐

c) 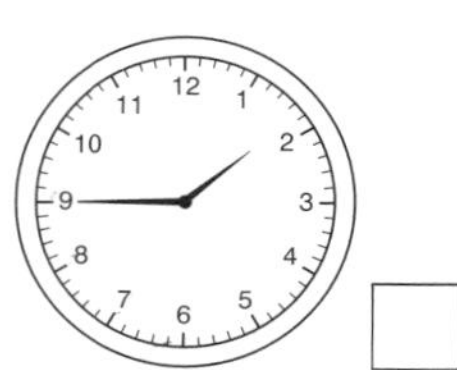 ☐

d) 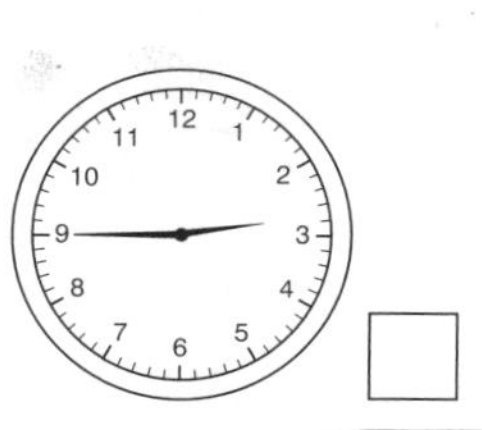 ☐

/1

Write the next two numbers in each **sequence**.

**27.** 30 26 22 18 14 ______ ______

**28.** 38 33 28 23 18 ______ ______

**29.** 43 41 39 37 35 ______ ______

**30.** 25 22 19 16 13 ______ ______

/4

/30

## PAPER 8

Write the missing < or > signs for each pair of numbers.

**1.** 34 ______ 51

**2.** 26 ______ 19

**3.** 40 ______ 38

**4.** 56 ______ 63

/4

Complete the grids.

**5.**

| + | 12 | 14 |
| --- | --- | --- |
| 7 | 19 | |
| 6 | | |

**6.**

| + | 6 | |
| --- | --- | --- |
| 9 | | 13 |
| | 14 | |

/2

Write these times.

**7.** 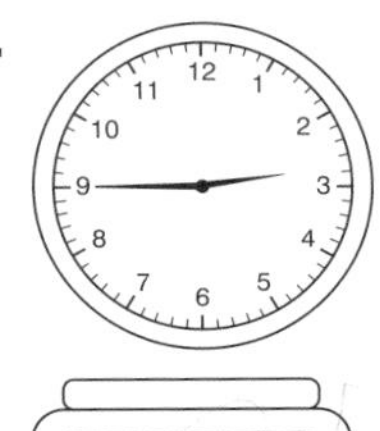

**8.** 

**9.** 

**10.** 

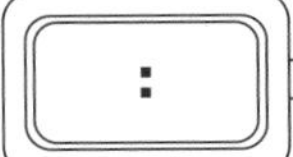

/4

**11.** Write 586 in words. ______________________________

/1

**12.** Write these six numbers in order, starting with the smallest.

225 306 251

340 299 260

______________________________

/1

**13.** How many days are there in December? ______

/1

**14.** Circle the rectangles.

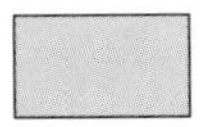 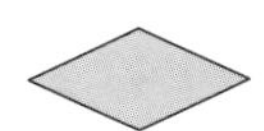  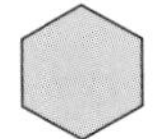 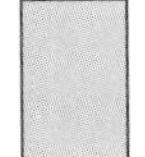 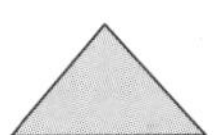

/1

Write the differences between each pair of numbers.

**15.** 38 43 difference → ______

**16.** 29 37 difference → ______

**17.** 46 60 difference → ______

**18.** 37 51 difference → ______

/4

**19.** Write these numbers on the Venn diagram.

35 17 28 43 30 25 32

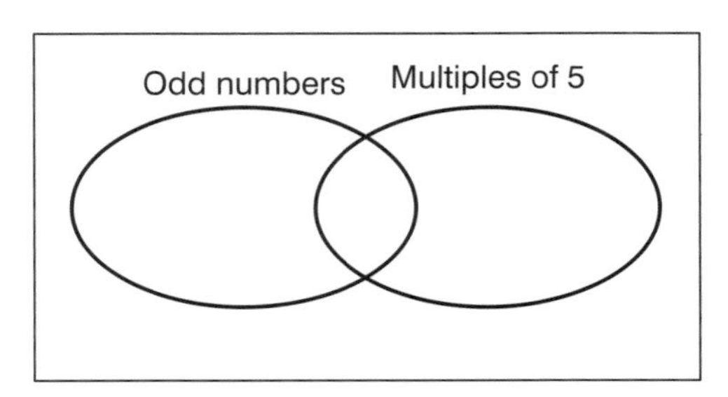


/1

How much water is in each jug?

**20.**

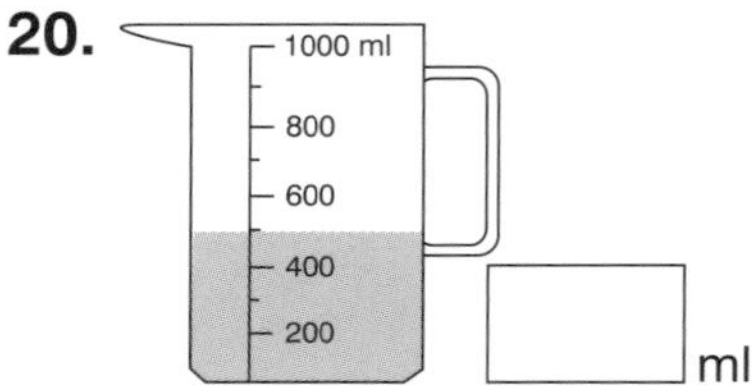


**21.**

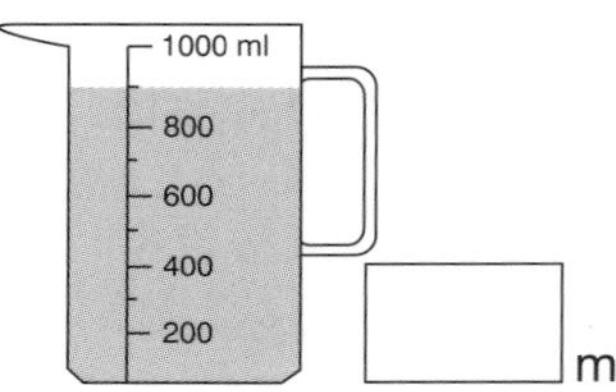


**22.**

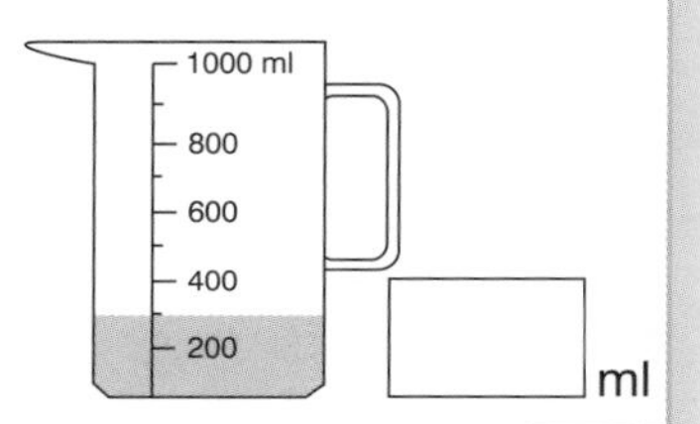


/3

Answer these.

**23.** What is 8 multiplied by 10? ________

**24.** What is 45 divided by 5? ________

**25.** What number multiplied by 4 gives the answer of 24? ________

/3

Answer these.

**26.** I'm thinking of a number. If I divide it by 2, the answer is 6.

What is my number? ________

**27.** I'm thinking of a number. If I multiply it by 2, the answer is 18.

What is my number? ________

28. I'm thinking of a number. If I divide it by 3, the answer is 3.

What is my number? ________

29. What is $\frac{1}{3}$ of 15? ________

30. What is $\frac{1}{4}$ of 24? ________

/5

/30

## PAPER 9

1. What is the next number in this sequence? Circle the answer.

127 137 147 157 ________

a) 177 b) 159 c) 167 d) 158

/1

Read and answer these questions.

2. What is the difference between 36 and 45? ________

3. How much greater is 81 than 75? ________

4. What number is 35 less than 50? ________

5. What is 80 take away 55? ________

/4

Look at the shaded area of each shape. Write **less than** $\frac{1}{2}$, **more than** $\frac{1}{2}$ or **equal to** $\frac{1}{2}$ for each fraction shaded.

6. 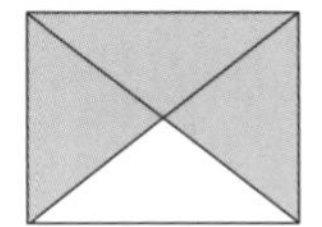

$\frac{3}{4}$ is ________

7. 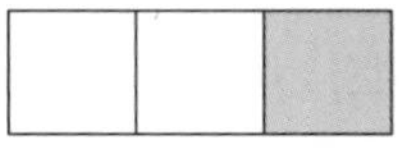

$\frac{1}{3}$ is ________

8. 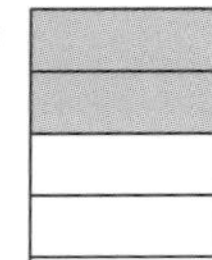

$\frac{2}{4}$ is ____________________

9. 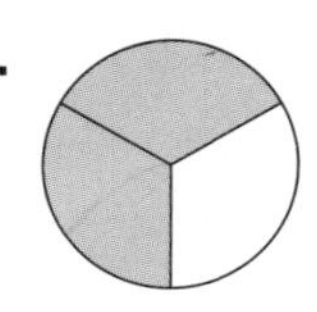

$\frac{2}{3}$ is ____________________

10. 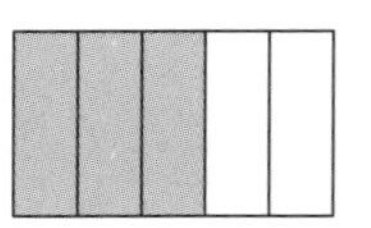

$\frac{3}{5}$ is ____________________

/5

11. Complete this multiplication grid.

| × | 5 | 2 | 8 |
|---|---|---|---|
| 3 | | 6 | |
| 10 | | | |
| 4 | | | |

/1

Write the names of these 3-D shapes.

12. 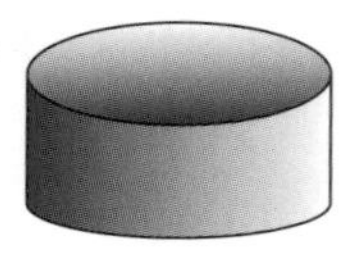

13. 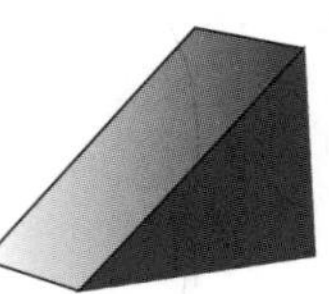

14. 

____________ ____________ ____________

/3

15. What is double twenty-six? ________

16. What is half of thirty-two? ________

17. A film starts at 5.15 p.m. and ends at 7.10 p.m.
How long is the film in hours and minutes?

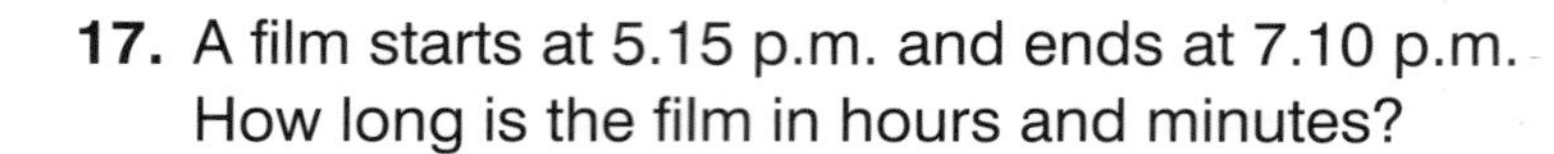

____________________

/2

**18.** What is the **area** shaded on this grid? ______ squares

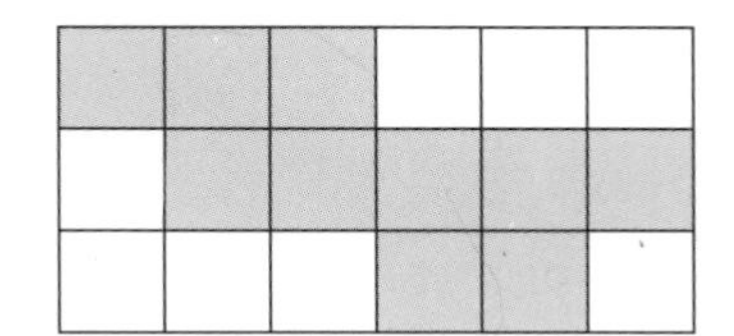

/1

Measure the length of each line.

**19.** Line A = ____________ cm

**20.** Line B = ____________ cm

**21.** Line C = ____________ cm

**22.** Line D = ____________ cm

/4

Write these numbers to make each sentence correct.

**23.** 575 495 ______ is less than ______

**24.** 419 407 ______ is less than ______

**25.** 612 621 ______ is less than ______

/3

Answer these.

**26.** 23 − 14 = ____________

**27.** 17 + 19 = ____________

**28.** 15 + 18 = ____________

**29.** 13 + 19 = ____________

**30.** 25 − 17 = ____________

/5

/30

## PAPER 10

The outside numbers total the centre number.
Write the missing numbers.

**1.**

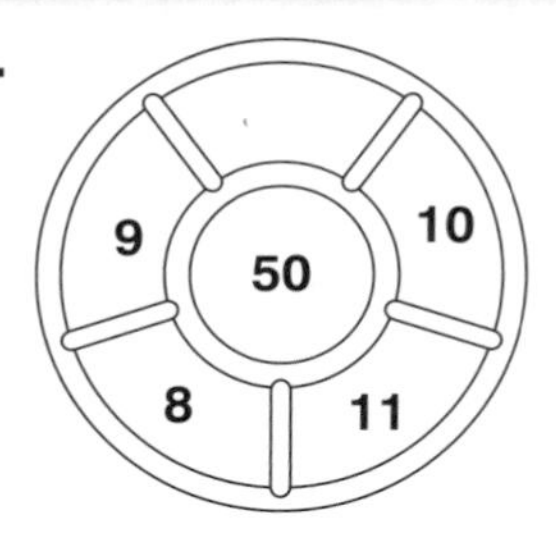


**2.**

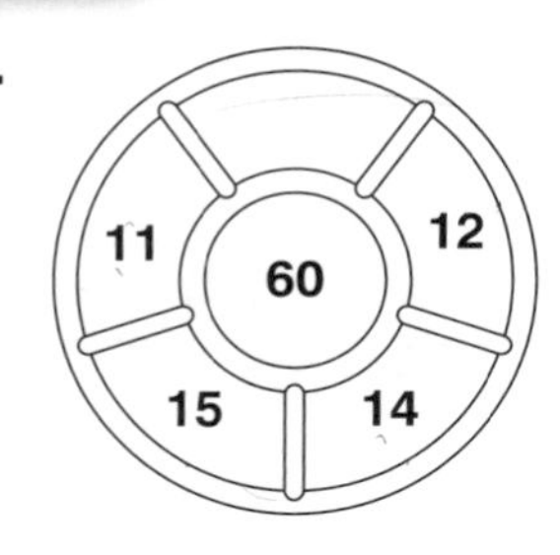


/2

**3.** True or false? These 3-D shapes are all prisms. Circle the answer.

True False

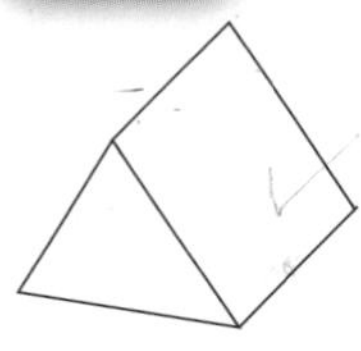

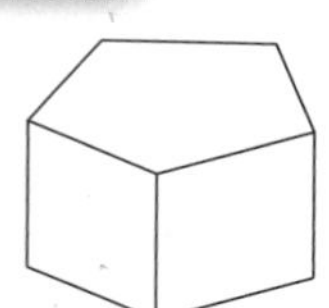

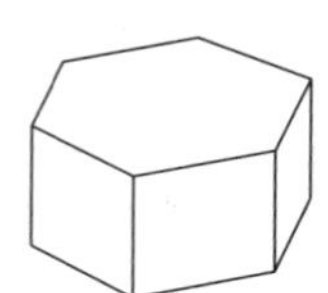

/1

Answer these.

**4.** A TV programme started at 4.15 and lasted for 30 minutes. At what time did it finish? ______________

**5.** Sam's train left the station at 9.45. The journey took an hour. What time did Sam arrive at the final station? ______________

**6.** Amy went to a swimming club at 5.15. She was there until 6.30. How long was she at the swimming club? ______________

**7.** A cake needs to be baked in the oven for 2 hours. If it needs to be taken out of the oven at 5.30, what time should it go in the oven? ______________

/4

Write the missing numbers for each machine.

**8.**

| IN | 26 | | 18 | | 15 |
|---|---|---|---|---|---|
| OUT | | 14 | | 22 | |

**9.**

| IN | 40 | | 36 | | 29 |
|---|---|---|---|---|---|
| OUT | | 25 | | 40 | |

/2

Tick the right angles in each shape. (Do not use a protractor)

**10.**

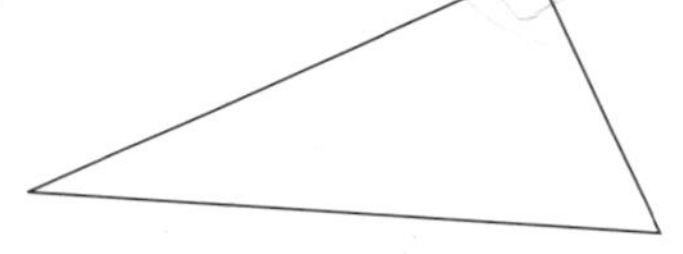

**11.**

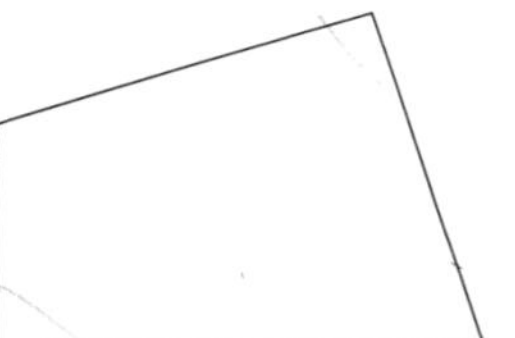

**12.**

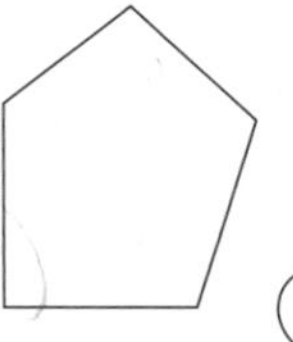

/3

This graph shows the number of people who visited a museum in a week.

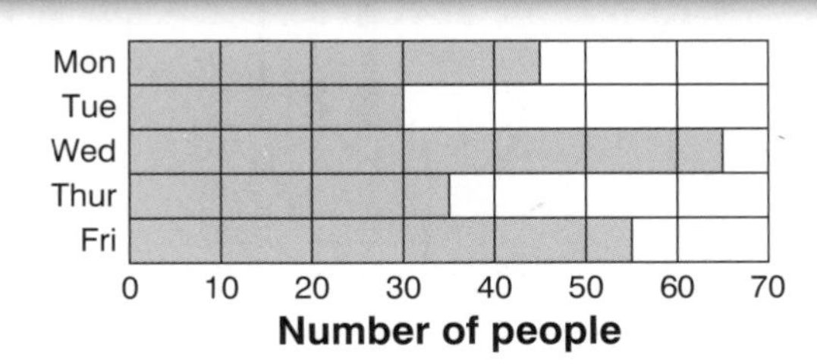


**13.** On which day did 35 people visit the museum? ______________

**14.** How many people visited the museum on Monday? ______________

**15.** On which day did 10 fewer people visit than on Wednesday? ______________

**16.** How many people visited in total on Monday and Tuesday? ______________

**17.** How many more people visited the museum on Friday than on Tuesday?

______________

/5

Write each group of numbers in order, starting with the smallest.

**18.** 89 91 72 95 ________________

**19.** 63 97 68 81 ________________

**20.** 40 88 32 69 ________________

**21.** 91 19 67 80 ________________

/4

Write the numbers shown on each abacus.

**22.**

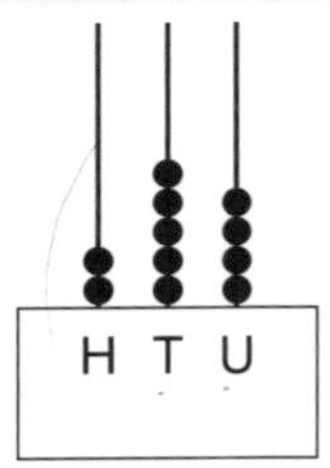


**23.**

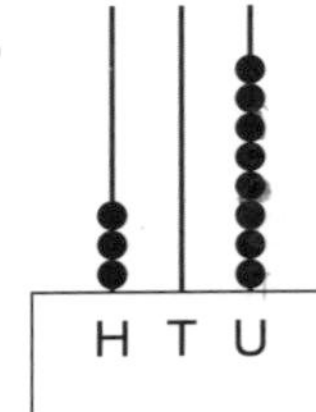


**24.**

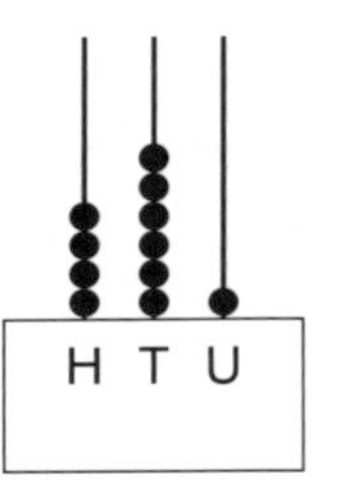


**25.**

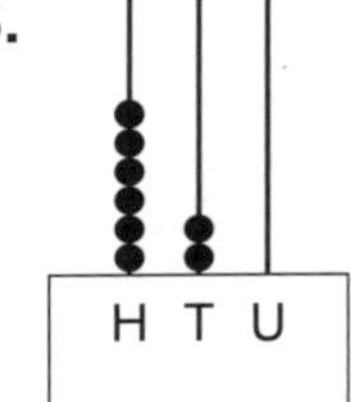


/4

Complete these.

**26.** 5000 g = _______ kg

**27.** 2000 ml = _______ l

**28.** 400 cm = _______ m

**29.** 3000 g = _______ kg

**30.** 6000 m = _______ km

/5

/30

# PAPER 11

Read and answer these.

1. What is the sum of 60 and 50? ______________

2. What is the total of 16 and 12? ______________

3. What is the difference between 24 and 19? ______________

4. Which number is 300 less than 800? ______________

5. Which number is 200 more than 600? ______________

6. What fraction of balls is circled? ______________

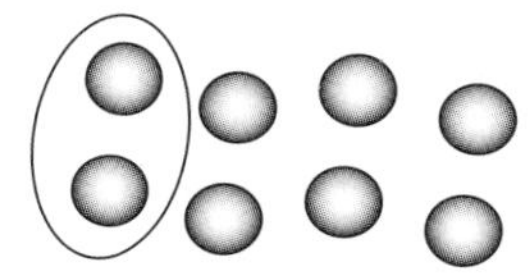

/6

7. Colour this grid to show a rectangle with an **area** of 12 squares.

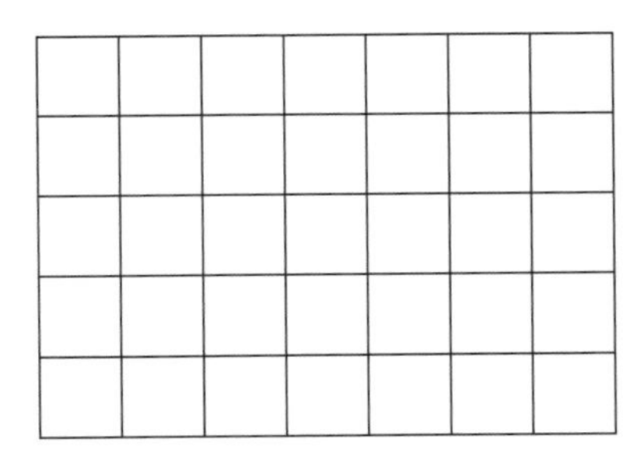

/1

How much liquid is in each jug?

8.

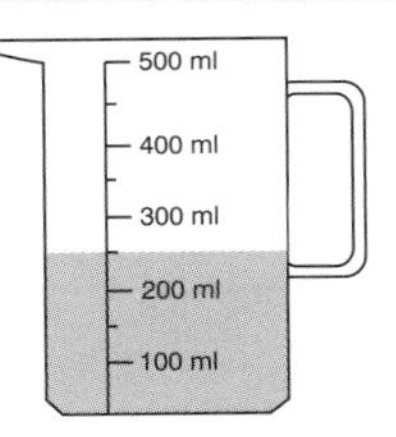


______________

9.

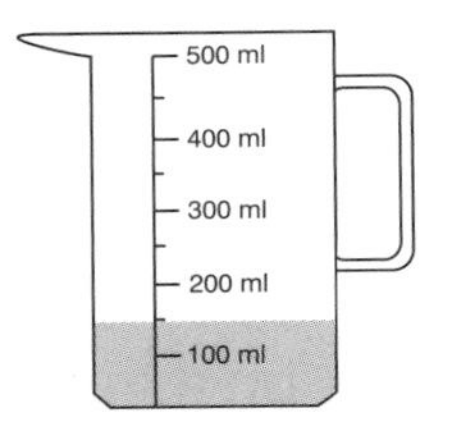


______________

/2

Write the number shown by each arrow.

**10.** **11.** **12.**

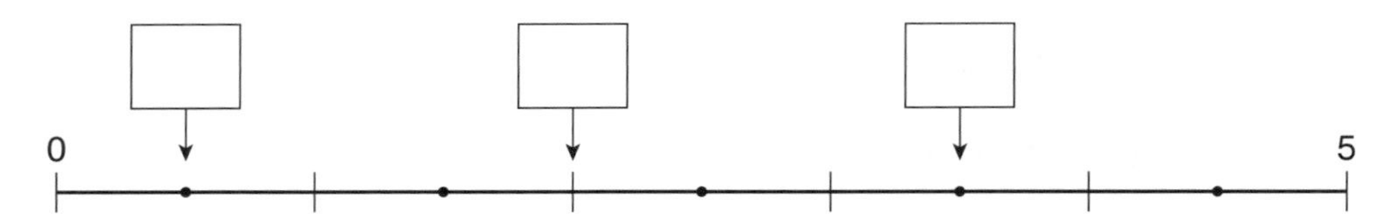


/3

Write these times.

**13.**

:

**14.**

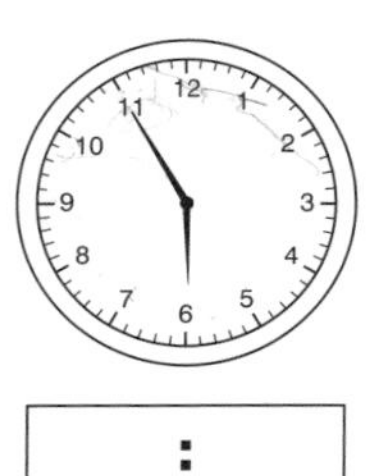

:

**15.**

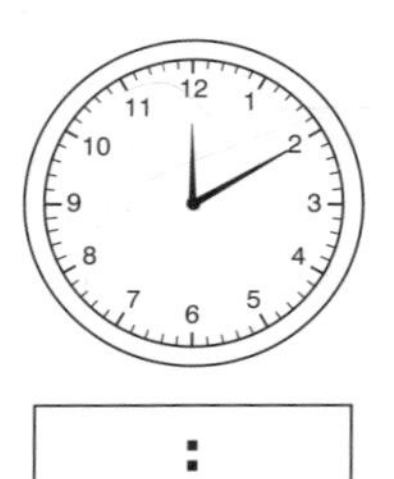

:

**16.**

:

/4

Answer these.

**17.** What is the name of a shape that has 5 straight sides? ______________

**18.** How many sides does a quadrilateral have? ______________

**19.** How many sides does a hexagon have? ______________

**20.** What is the name of a shape that has 3 straight sides? ______________

/4

Write the totals.

**21.** £3.20 + 45p = £__________

**22.** £1.90 + 50p = £__________

**23.** £4.70 + 35p = £__________

**24.** £3.60 + 60p = £__________

**25.** There are 18 cakes on a tray ready to be put in packets.
Each packet holds 4 cakes.
How many packets are needed to hold all 18 cakes?

______________

/5

Write the next two numbers in each sequence.

26. 175 176 177 178 ______ ______

27. 384 383 382 381 ______ ______

28. 272 282 292 302 ______ ______

29. 319 309 299 289 ______ ______

30. I'm thinking of a number. If I take away 9 from it, the answer is 11.
What is my number?

______

/5

**/30**

## PAPER 12

Use these numbers to answer the questions.

10 11 14 9 15

1. What is the largest **even** total made by adding two numbers? ______
2. What is the smallest **odd** total made by adding two numbers? ______
3. Which two numbers total 20? ______
4. Which two numbers total 25? ______

/4

Draw the line of symmetry on each of these shapes.

5.

6.

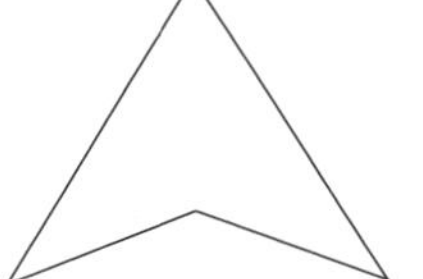

7.

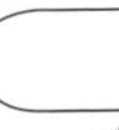

/3

Answer these.

**8.** $5 \times 6 =$ ☐

**9.** $3 \times 8 =$ ☐

**10.** $9 \times 2 =$ ☐

**11.** $7 \times 10 =$ ☐

**12.** Tick the shape that has one-third shaded.

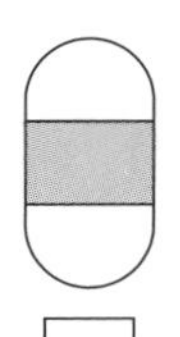
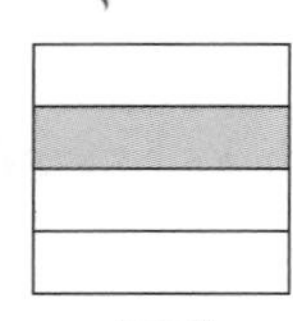
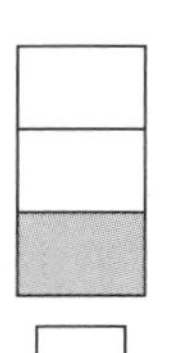
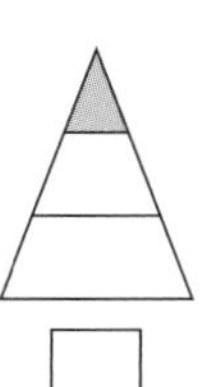

☐ ☐ ☐ ☐

/1

**13.** I buy five tickets costing 50p each.
How much change will I get from £5?

______________

Choose from these shapes to answer the questions.

**prism cone cube cylinder**

**14.** Name a shape with at least one curved **face**. ______________

**15.** Name a shape that has only flat faces. ______________

**16.** Name a shape that has 6 identical square faces. ______________

**17.** Name a shape that has a least one flat circle face. ______________

/4

Answer these.

**18.** Total 15 and 7. ______________

**19.** Add 6 to 19. ______________

**20.** Increase 18 by 4. ______________

/3

# Answer booklet: Success Assessment Papers Maths 7–8 years

**Paper 1**

**1.** b) 80 **2.** 22 24
**3.** 25 23 **4.** 51 53
**5.** 54 52 **6.** c)

**7.** $\frac{1}{2}$ **8.** $\frac{1}{3}$
**9.** $\frac{1}{4}$ **10.** 19
**11.** 18 **12.** 19
**13.** 16 **14.**
**15.** 4 **16.** 7
**17.** 5 **18.** 5
**19.** 7 cm **20.** 4 cm
**21.** 8 cm
**22.**

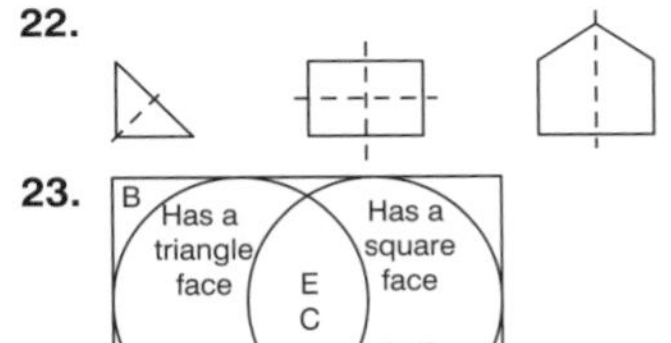

**23.**

B
Has a triangle face
Has a square face
E
C
A
D, F

**24.** 18 **25.** 14
**26.** 8 **27.** ninety-one
**28.** 56 **29.** eighty-three
**30.** 74

**Paper 2**

**1.** Wednesday **2.** Saturday
**3.** Tuesday **4.** Monday
**5.**
**6.**

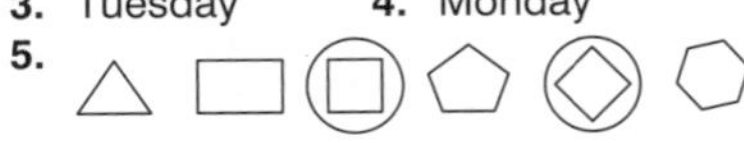

**7.**

**8.**

**9.** £1.65 **10.** 11 kg
**11.** 12 **12.** 15
**13.** 80 **14.** 18
**15.** 60 75 90 105 120 135 150
**16.** 8 **17.** 15
**18.** 6 **19.** 21
**20.** 4 **21.** 6
**22.** eighty-seven **23.** cube
**24.** triangular prism
**25.** cone **26.** 80
**27.** 20 **28.** 40
**29.** 160 **30.** 150

**Paper 3**

**1.** 77 **2.** 99
**3.** 94 **4.** 85
**5.** **6.** 5
**7.** 3 **8.** 2
**9.** 21
**10.** Any four squares coloured on grid.
**11.** 18 27 36 45
**12.** box C **13.** 60p
**14.** 55, 27, 91, 85
**15.** 74 **16.** 91
**17.** 86 **18.** 6
**19.** 7.00 **20.** 6
**21.** 9 **22.** 8 cm
**23.** 9 cm **24.** 1 cm
**25.** 50 **26.** 80
**27.** 37 42 **28.** 25 27
**29.** 36 39 **30.** 34 38

**Paper 4**

**1.** £1.90 **2.** £3.50
**3.** £2.35 **4.** £2.65
**5.** 31 **6.** 36
**7.** 26 **8.** 35
**9.**

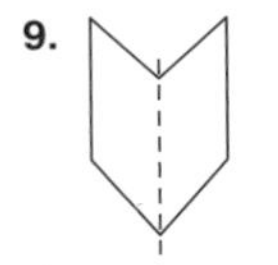

**10.**

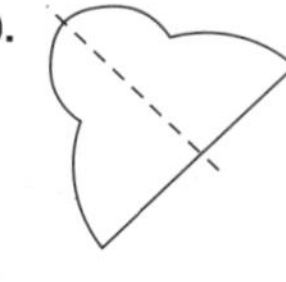

**11.**

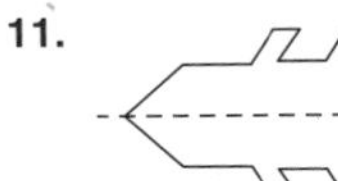

**12.**

| IN | 15 | 13 | 20 | 16 | 18 |
|---|---|---|---|---|---|
| OUT | 6 | 4 | 11 | 7 | 9 |

**13.** 8:15 **14.** 6:45
**15.** 3:45 **16.** 8:15
**17.** 2 litres
**18.** 11 15 19 22
**19.** 17 18 21 23
**20.** £3 **21.** £2.50
**22.** triangle **23.** square
**24.** pentagon **25.** hexagon
**26.** 50 92 28 76
**27.** 18 **28.** 12
**29.** 50 **30.** 18

**Paper 5**

**1.** 15 squares
**2.** $165 = 100 + 60 + 5$
**3.** $219 = 200 + 10 + 9$
**4.** $456 = 400 + 50 + 6$
**5.** $3 + 9 = 12$ **6.** $11 - 6 > 4$
**7.** $5 + 6 < 13$ **8.** $10 - 4 < 11$
**9.** $9 + 6 = 15$ **10.** $13 - 4 > 6$
**11.** $3 \times 4 = 2 \times 6$
$9 \times 2 = 6 \times 3$
$1 \times 6 = 3 \times 2$
**12.** 7 **13.** 3
**14.** 5 **15.** 27 47
**16.** 17 26 **17.** 8
**18.** tennis **19.** 10
**20.**

| IN | 18 | 20 | 25 | 16 | 23 |
|---|---|---|---|---|---|
| OUT | 7 | 9 | 14 | 5 | 12 |

**21.** 105 **22.** 150
**23.** 115 **24.** cuboid
**25.** cylinder **26.** 5 cm
**27.** 8 cm **28.** 3 cm
**29.** 3 **30.** 6

**Paper 6**

**1.** 34 **2.** 35
**3.**

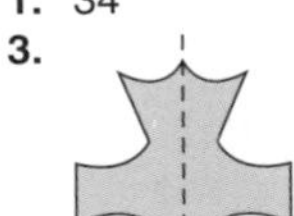

**4.**

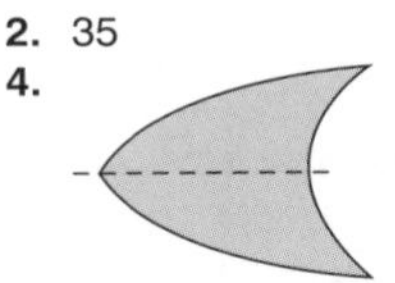

**5.** 105 **6.** 128
**7.** 124 **8.** 133
**9.** 142 **10.** 7 cm
**11.** 5 cm **12.** 8 cm
**13.** 58 **14.** 72
**15.** 35 and 60 **16.** 12 and 23
**17.** 45 minutes **18.** 15 minutes
**19.** 30 minutes **20.** 10
**21.** 2 **22.** 5
**23.** 20 **24.** £2.70
**25.** £1.72 **26.** 9
**27.** 11 **28.** 4
**29.** 5 **30.** 4

**Paper 7**

**1.** $215 = 200 + 10 + 5$
**2.** $193 = 100 + 90 + 3$
**3.** $262 = 200 + 60 + 2$

4. 517 = 500 + 10 + 7
5. 321 = 300 + 20 + 1
6. 486 = 400 + 80 + 6
7. rectangle
8. right-angled triangle
9. pentagon 10. semi-circle
11. 25p 12. 11p
13. 39p 14. 17p
15. East 16. North
17. North 18. West
19. 15 20.

21.  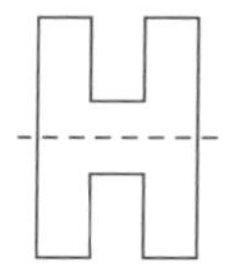 22. 800 cm

23. 650 cm 24. 400
25. 701 26. c) 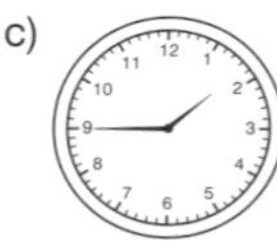
27. 10 6 28. 13 8
29. 33 31 30. 10 7

**Paper 8**

1. 34 < 51 2. 26 > 19
3. 40 > 38 4. 56 < 63

5.

| + | 12 | 14 |
|---|---|---|
| 7 | 19 | 21 |
| 6 | 18 | 20 |

6.

| + | 6 | 4 |
|---|---|---|
| 9 | 15 | 13 |
| 8 | 14 | 12 |

7. 2:45 8. 11:15
9. 1:45 10. 10:15
11. five hundred and eighty-six
12. 225 251 260 299 306 340
13. 31
14. 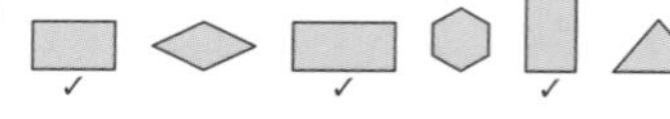
15. 5 16. 8
17. 14 18. 14
19. 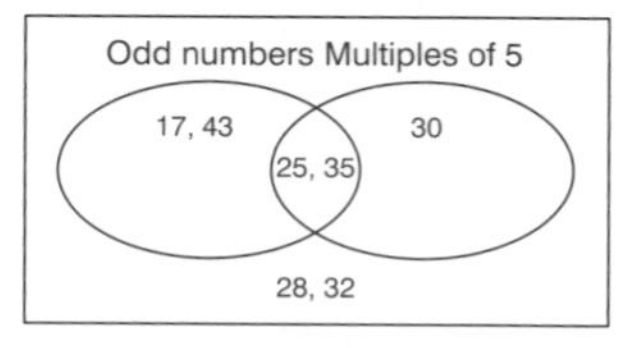

20. 500 ml 21. 900 ml
22. 300 ml 23. 80
24. 9 25. 6
26. 12 27. 9
28. 9 29. 5
30. 6

**Paper 9**

1. 167 2. 9
3. 6 4. 15
5. 25 6. more than $\frac{1}{2}$
7. less than $\frac{1}{2}$ 8. equal to $\frac{1}{2}$
9. more than $\frac{1}{2}$
10. more than $\frac{1}{2}$
11.

| × | 5 | 2 | 8 |
|---|---|---|---|
| 3 | 15 | 6 | 24 |
| 10 | 50 | 20 | 80 |
| 4 | 20 | 8 | 32 |

12. cylinder
13. triangular prism
14. sphere 15. 52
16. 16
17. 1 hour 55 minutes
18. 10 19. 3 cm
20. 7 cm 21. 6 cm
22. 8 cm
23. 495 is less than 575
24. 407 is less than 419
25. 612 is less than 621
26. 9 27. 36
28. 33 29. 32
30. 8

**Paper 10**

1. 12 2. 8
3. true 4. 4.45
5. 10.45
6. 1 hour 15 minutes
7. 3.30
8.

| IN | 26 | 9 | 18 | 17 | 15 |
|---|---|---|---|---|---|
| OUT | 31 | 14 | 23 | 22 | 20 |

9.

| IN | 40 | 13 | 36 | 28 | 29 |
|---|---|---|---|---|---|
| OUT | 52 | 25 | 48 | 40 | 41 |

10. 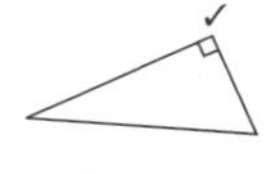 11. 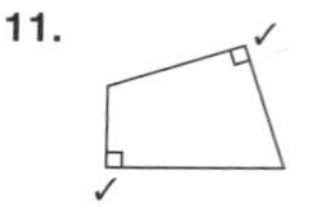
12. 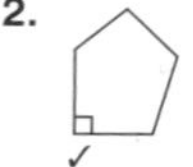
13. Thursday 14. 45 people
15. Friday 16. 75 people
17. 25 people
18. 72 89 91 95
19. 63 68 81 97
20. 32 40 69 88
21. 19 67 80 91
22. 254 23. 308
24. 461 25. 620

26. 5000 g = 5 kg
27. 2000 ml = 2 l
28. 400 cm = 4 m
29. 3000 g = 3 kg
30. 6000 m = 6 km

**Paper 11**

1. 110 2. 28
3. 5 4. 500
5. 800 6. $\frac{1}{4}$ or $\frac{2}{8}$
7. Check rectangle has an area of 12 squares.
8. 250 ml 9. 300 ml
10. $\frac{1}{2}$ 11. 2
12. $3\frac{1}{2}$ 13. 1.15
14. 5.55 15. 12.10
16. 9.05 17. pentagon
18. 4 19. 6
20. triangle 21. £3.65
22. £2.40 23. £5.05
24. £4.20 25. 5 packets
26. 179 180 27. 380 379
28. 312 322 29. 279 269
30. 20

**Paper 12**

1. 26 2. 19
3. 11 and 9
4. 10 and 15 or 11 and 14
5. 6. 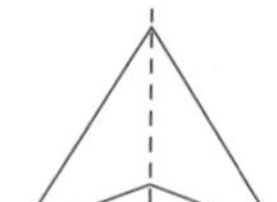
7. 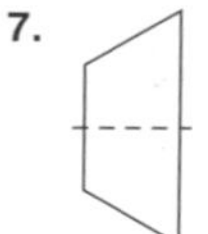
8. 30 9. 24
10. 18 11. 70
12. 13. £2.50
14. cone or cylinder
15. prism or cube
16. cube
17. cone or cylinder
18. 22 19. 25
20. 22 21. 27
22. 31 23. $7\frac{1}{2}$ cm
24. South 25. 2.20
26. 4.45 27. 10.35
28. 7.25
29. 60, 30 and 10
30. 13 + 4 < 14 + 5

## Paper 13

1. $\frac{1}{2}$
2. $1\frac{1}{2}$
3. $2\frac{1}{4}$
4. $3\frac{1}{4}$
5. $4\frac{3}{4}$
6. 90
7. 70
8. 110
9. 140
10. +11
11. 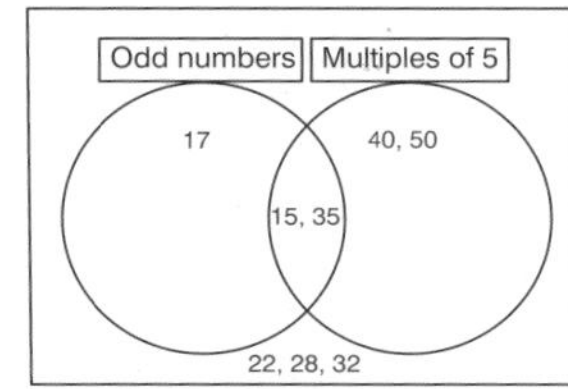

12. 657 658 659 660 661 662 663
13. Check line is between 4 and 6 cm long.
14. Check line is between 8 and 10 cm long.
15. Check line is between 2 and 4 cm long.
16. C
17. B
18. A and E
19. D
20. 560 is less than 650
21. 419 is less than 431
22. 521 is less than 541
23. 306 is less than 316
24. July
25. 4th July
26. Friday
27. 4
28. 31
29. 17th
30. Saturday

## Paper 14

1. 3
2. 6
3. 2
4. 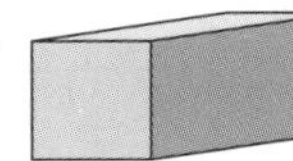
5. +4
6. $762 = 700 + 60 + 2$
7. $529 = 500 + 20 + 9$
8. $614 = 600 + 10 + 4$
9. $931 = 900 + 30 + 1$
10. 604
11. 2
12. 4
13. 4
14. 5
15. 10
16. $\frac{2}{5}$
17. $\frac{3}{8}$
18. $\frac{3}{4}$
19. $\frac{5}{6}$
20. 16
21. 14
22. 21
23. 41
24. £3.12
25. £5.09
26. £4.88
27. £6.20
28. non-symmetrical
29. symmetrical
30. non-symmetrical

## Paper 15

1. 4
2. 8
3. 7
4. 9
5. 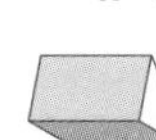
6. South
7. North
8. East
9. West
10. $<$
11. $>$
12. $<$
13. $>$
14. $>$
15. 170
16. C
17. A
18. D
19. B
20. A
21. C
22. 250 ml
23. 650 ml
24. 85 + 15, 74 + 26, 75 + 25, 64 + 36
25.   
6:05
26. 
12:20
27. 
3:10
28.  
9:55
29. 30
30. 3

## Paper 16

1. 30
2. 24
3. 28
4. 40
5. square-based pyramid
6. triangular prism
7. cuboid
8.

| × | 3 | 10 | 6 |
|---|---|---|---|
| 5 | 15 | 50 | 30 |
| 4 | 12 | 40 | 24 |
| 2 | 6 | 20 | 12 |

9. 35
10. 20
11. 18
12. 40
13. 8
14. 24
15. 14 days
16. 2 hours
17. 180 minutes
18. 3 weeks
19. 24 months
20. 2 days
21. £1, 50p, 20p, 10p
22. £1, 20p, 5p 1p
23. Supermarket
24. Jeweller's
25. 40 people
26. Chemist's
27. 2
28. 8
29. 10
30. 3

## Paper 17

1.

| IN | 7 | 3 | 18 | 20 | 35 |
|---|---|---|---|---|---|
| OUT | 700 | 300 | 1800 | 2000 | 3500 |

2. 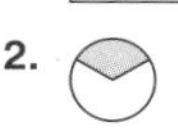 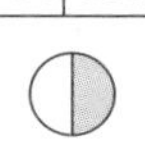 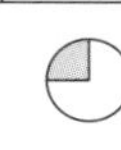 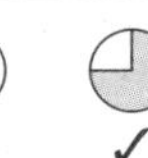 ✓
3. $\frac{2}{3} = \frac{4}{6}$
4. $35 + 6 < 50 - 8$
5. $62 - 4 > 49 + 6$
6. 10 squares
7. 473
8. 812
9. 380
10. 540, 440
11. 183, 185
12. 320, 310
13. 212, 222
14. 510, 508
15. 
16. 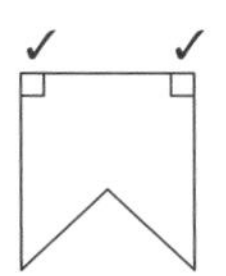
17. 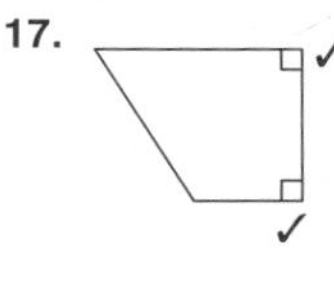
18. 850 ml
19. 450 ml
20. 50 ml
21. 78
22. 70
23. 61
24. 81
25. 1
26. 3
27. $4\frac{1}{2}$
28. $6\frac{1}{2}$
29. $9\frac{1}{2}$
30. 6 tubes

## Paper 18

1. 2000 g
2. 500 g
3. 5000 g
4. 250 g
5. 750 g
6. 82
7. 72
8. 85
9. 84
10. sphere
11. cone
12. cylinder

13–15. 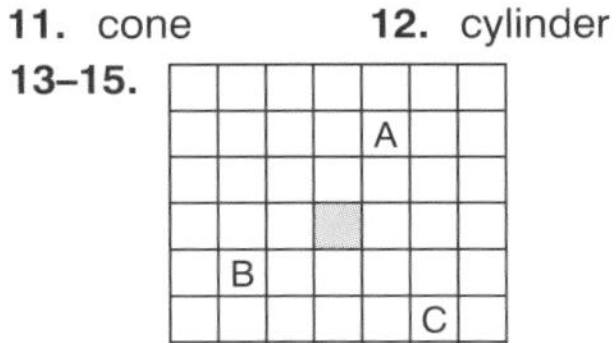


16. 16
17. 26
18. 18
19. 11
20. 9
21. 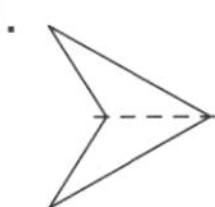
22. 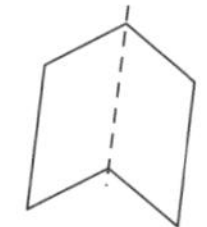
23. 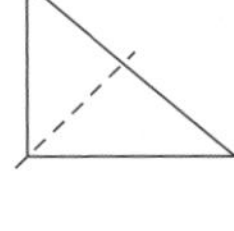
24. 10, 15, 25

25. Line A = $5\frac{1}{2}$ cm

26. Line B = $3\frac{1}{2}$ cm

27. Line C = 8 cm

28. Line D = $6\frac{1}{2}$ cm

29. 350 30. 280

**Paper 19**

1. 5 2. 7
3. 6 4. 23
5. 1014 6. 24
7. 70, 75 8. 70, 60
9. 40, 42 10. 74, 77
11. ✓ ✓ ✓
12. 911 13. 32
14. 13 15. 49
16. 37 17. 18
18. 16
19. 400 + 80 + 6
20. 900 + 40 + 5
21. 800 + 90 + 4
22. 190 23.

| | A D E |
|---|---|
| B C | F |

24. 1 m 10 cm 25. 250 cm
26. 2 m 20 cm 27. 340 cm
28. 5 m 80 cm 29. £8
30. £6

**Paper 20**

1. 8.15 2. 7.10
3. 8.05 4. 12.55
5. 35 6. 8
7. 6 8. 9
9. 4 10. 9
11. Mirror line 12.

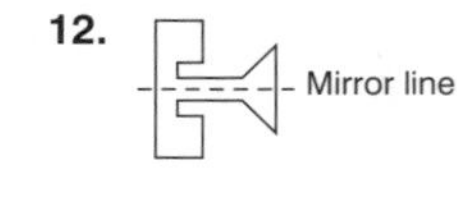


13. 51 14. 124
15. 54 16. 115
17. 500 18. 600
19. 800 20. 900
21. 528 586 619 691
22. 937 940 941 947
23. 804 840 848 884
24.

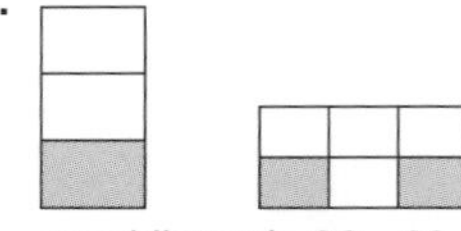

25. quadrilateral 26. 90
27. 110 28. 72
29. 123
30. Check that a right-angled triangle has been drawn.

**Paper 21**

1.

| 2 | 4 | 5 | 7 | 8 |
|---|---|---|---|---|
| 4 | 8 | 10 | 14 | 16 |

2.

| 3 | 5 | 6 | 7 | 9 |
|---|---|---|---|---|
| 15 | 25 | 30 | 35 | 45 |

3. 608 4. 5.25
5. 7.50 6. 4.30
7. 11.15 8. 2.55
9. 10.40 10. 8
11. 4 12. 16
13. 29 14. 54
15. 26
16. 100 120 140 160 180 200 220
17. 140 130 120 110 100 90 80
18. 108 106 104 102 100 98 96
19. 25 30 35 40 45 50 55
20.

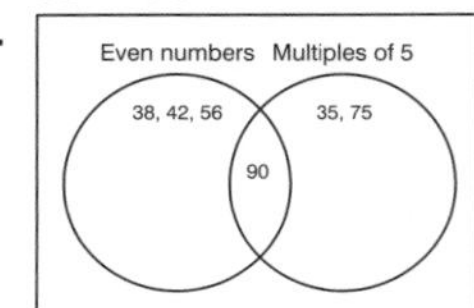


21. 60 ml 22. 30 ml
23. 85 ml 24.

25. 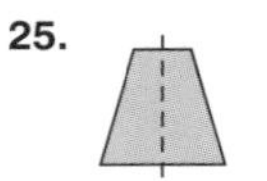 26. 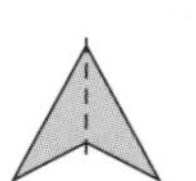

27.   28. 110

29. 700 30. 1000

**Paper 22**

1. 16 2. 12
3. 15 4. 18
5. 371 6. East
7. North 8. West
9. South
10. nine hundred and eighty-four
11. 475
12. seven hundred and eighty-nine
13. 819 14. 600 ml
15. 800 ml 16. 300 ml
17. hexagon 18. 6 cm
19. $5\frac{1}{2}$ cm 20. 8 cm
21. 55 22. 260
23. 150 24. 115
25.

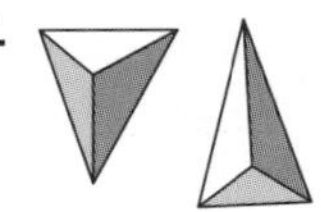

26. 2 × 6, 3 × 4, 6 × 2 , 4 × 3
27. 9 × 2, 3 × 6, 2 × 9, 6 × 3
28. 9 29. 23
30. 80

**Paper 23**

1. 9 + 9 < 21 2. 15 − 8 = 7
3. 14 + 7 > 19 4. 23 − 4 > 17
5. 18 + 11 < 33 6. 25 − 9 = 16
7.

| 2 | 4 | 5 | 7 | 10 |
|---|---|---|---|---|
| 6 | 12 | 15 | 21 | 30 |

8.

| 4 | 5 | 7 | 9 | 10 |
|---|---|---|---|---|
| 16 | 20 | 28 | 36 | 40 |

9. 537 10. 306
11. 920 12. 10
13. 8 14. 5
15. 7 16. 5
17. 5000 ml 18. 6 m
19. 7 km 20. 3 kg
21. cuboid 22. 2
23. 4 24. 120 minutes
25. 14 days 26. 24 months
27. Friday 28. October
29. cone 30. $\frac{2}{4}, \frac{3}{6}, \frac{5}{10}$

**Paper 24**

1. 5 2. 14
3. 9 4. 9
5. 33 6. 26
7. 26 8. 23
9–11.

| | A | | | | | |
|---|---|---|---|---|---|---|
| | | | | | | |
| | | | | | | |
| | | | ■ | | | |
| | | | | | | |
| | | | | | | B |
| C | | | | | | |

12. 604 608 643 650
13. 235 319 325 349
14. 506 536 555 560
15. 407 474 740 744
16. 2 × 12, 4 × 6, 8 × 3, 24 × 1
17. 5 × 6, 3 × 10, 15 × 2, 30 × 1
18. Ben and Claire
19. bag 20. Alex
21. brown 22. Ben
23. Alex and Donna
24. sphere
25. triangular prism
26. pyramid 27. cone
28. 25 minutes 29. 30 minutes
30. 50 minutes

**21.** Add together 8, 9 and 10. ______________

**22.** Total 7, 11 and 13. ______________

**23.** What length is the line to the nearest $\frac{1}{2}$ centimetre? ______________

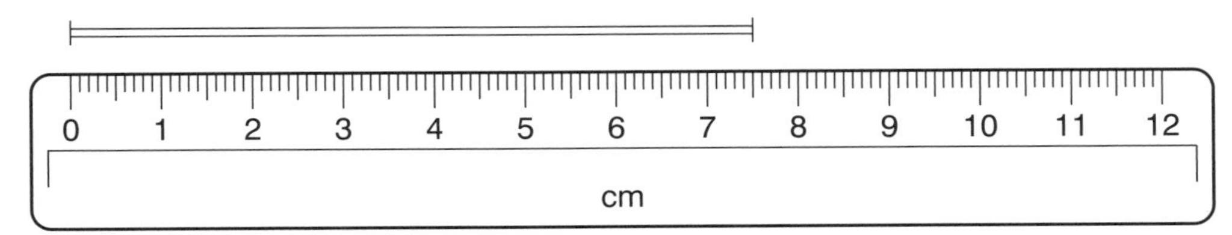


**24.** I am facing East and I make a quarter turn **clockwise**. In which direction am I now facing?

______________

/4

Write these times.

**25.** 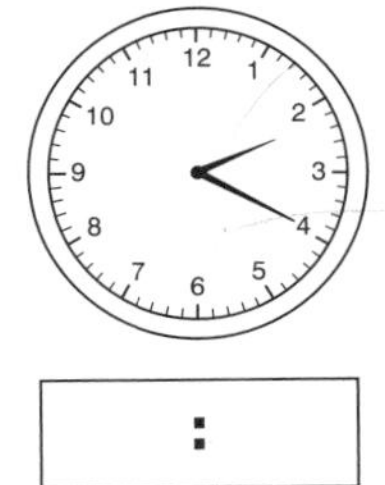

[ : ]

**26.** 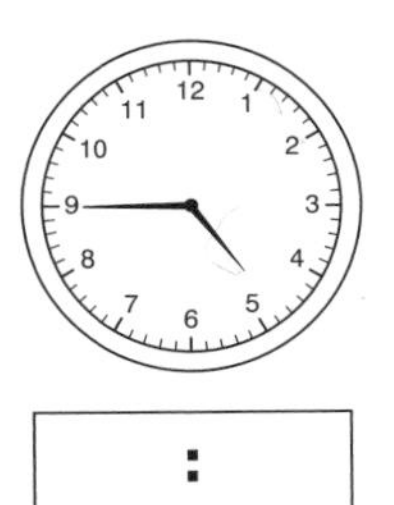

[ : ]

**27.** 

[ : ]

**28.** 

[ : ]

**29.** Circle three numbers that add up to make 100.

20 40 60 30 10

**30.** Choose a sign to make this correct.

13 + 4 _____ 14 + 5

/6

/30

# PAPER 13

Write these numbers on the number line.

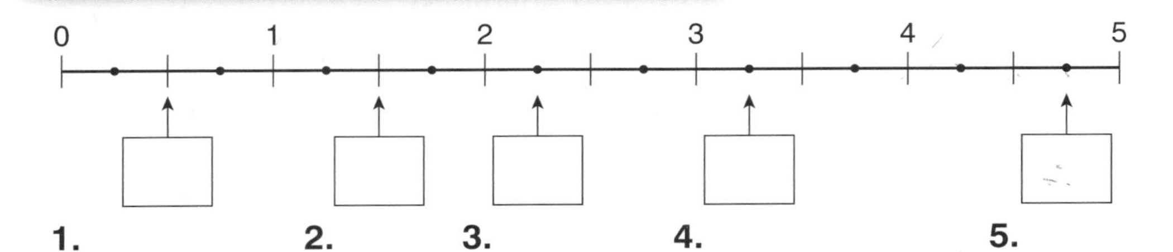


**1.** **2.** **3.** **4.** **5.**

/5

**Round** these numbers to the nearest ten.

**6.** 87 → ____________

**7.** 65 → ____________

**8.** 114 → ____________

**9.** 136 → ____________

/4

**10.** What is this machine doing to the numbers in the grid? Write the answer on the box in the machine.

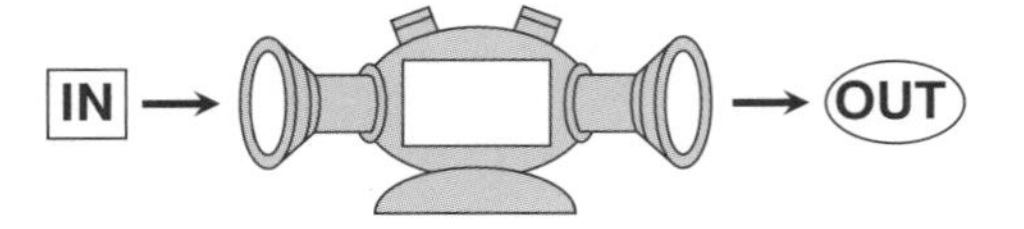


| IN | 0 | 3 | 5 | 6 | 8 |
|---|---|---|---|---|---|
| OUT | 11 | 14 | 16 | 17 | 19 |

/1

**11.** Write the following numbers in the correct place on this Venn diagram.

17 22 35 40 28 15 50 32

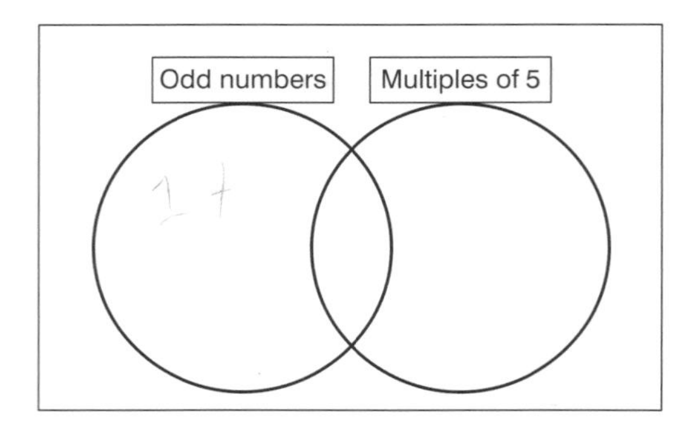


/1

**12.** Write the missing numbers in the correct order.

(660) (659) (662) (658)

657 ______ ______ ______ 661 ______ 663

/1

Do not use a ruler. In each box, draw a straight line that you **estimate** to be about the correct length.

**13.** 5 cm [ ]

**14.** 9 cm [ ]

**15.** 3 cm [ ]

/3

Look at these angles.

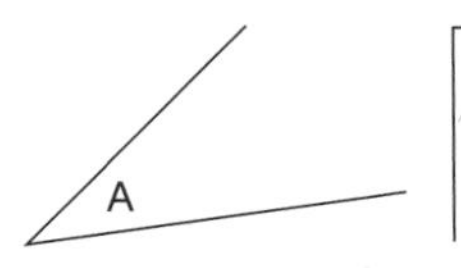
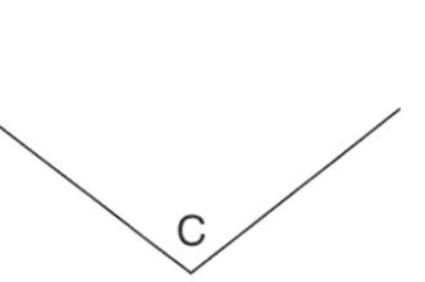
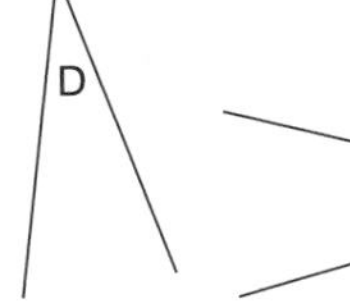


**16.** Which is the largest angle? ______

**17.** Which is a right angle? ______

**18.** Which two angles are the same size? ______

**19.** Which is the smallest angle? ______

/4

Write these numbers to make each sentence correct.

**20.** 560 650 ______ is less than ______

**21.** 419 431 ______ is less than ______

**22.** 541 521 ______ is less than ______

**23.** 306 316 ______ is less than ______

/4

Answer the questions about this calendar.

**JULY**

| Mon | Tues | Weds | Thurs | Fri | Sat | Sun |
|---|---|---|---|---|---|---|
| | | 1 | 2 | 3 | 4 | 5 |
| 6 | 7 | 8 | 9 | 10 | 11 | 12 |
| 13 | 14 | 15 | 16 | 17 | 18 | 19 |
| 20 | 21 | 22 | 23 | 24 | 25 | 26 |
| 27 | 28 | 29 | 30 | 31 | | |

**24.** What month does the calendar show? ____________________

**25.** What date is the first Saturday? ____________________

**26.** On which day is it the 24th July? ____________________

**27.** How many Tuesdays are there in the month in total? ____________________

**28.** How many days are there in July in total? ____________________

**29.** What date in July is the third Friday? ____________________

**30.** What day of the week is the 1st August? ____________________

/7

/30

## PAPER 14

Complete these.

**1.** $8 \times \_\_\_\_ = 24$

**2.** $\_\_\_\_ \times 4 = 24$

**3.** $12 \times \_\_\_\_ = 24$

/3

4. Which of these 3-D shapes has 6 **faces**? Tick the shape.

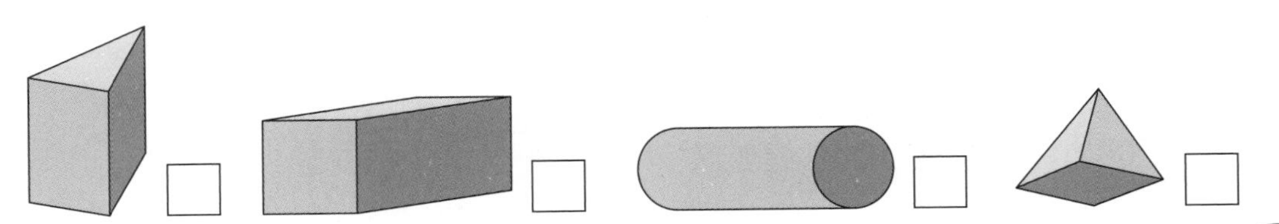

/1

5. What is this machine doing to the numbers? Write the answer in the box on the machine.

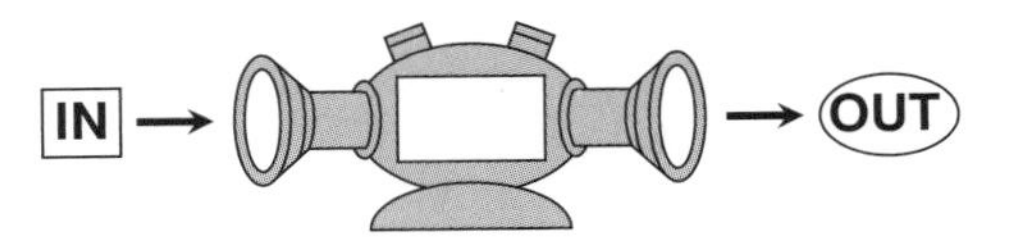


| IN | 7 | 8 | 10 | 12 | 15 |
|---|---|---|---|---|---|
| OUT | 11 | 12 | 14 | 16 | 19 |

/1

Write the missing numbers.

6. 762 = ______ + 60 + ______

7. 529 = ______ + 20 + ______

8. 614 = ______ + 10 + ______

9. 931 = ______ + 30 + ______

/4

10. Choose the missing number

554 < ______

468 399 604 509

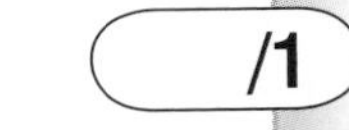


11. How many 500 ml bottles will fill a 1 litre jug? ______

12. How many 500 ml bottles will fill a 2 litre jug? ______

13. How many 250 ml cups will fill a 1 litre jug? ______

14. How many 100 ml spoons will fill a 500 ml water bottle? ______

15. How many 100 ml spoons will fill a 1 litre jug? ______

/5

Level 3

What fraction of each of these shapes is shaded?

**16.** 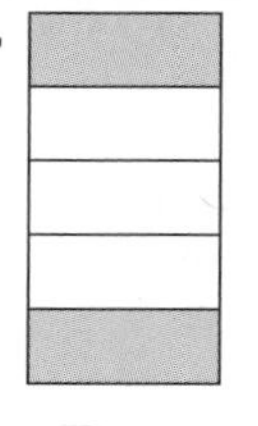 ________

**17.** 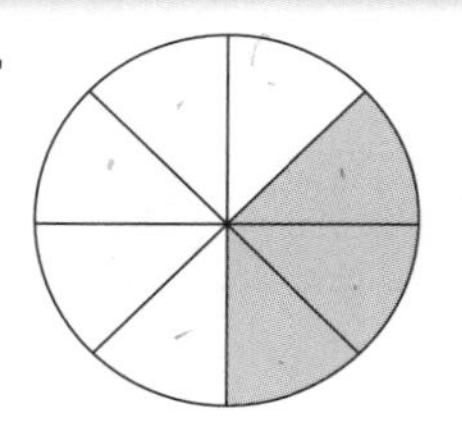 ________

**18.** 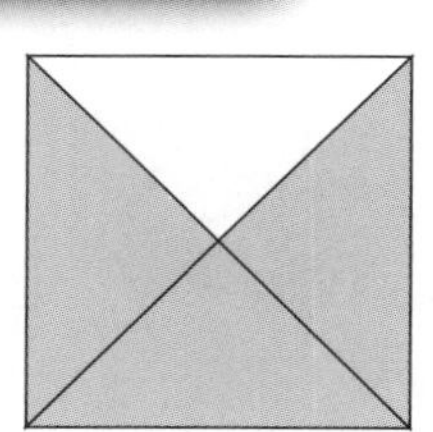 ________

**19.** 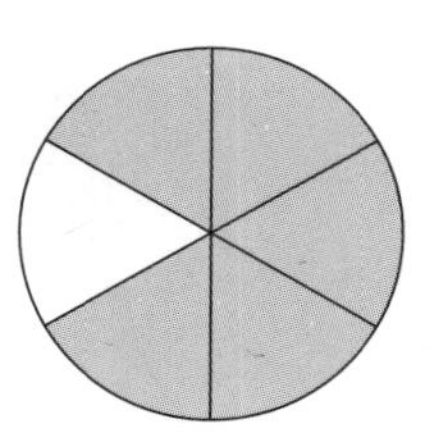 ________

/4

Answer these.

**20.** 48 − 32 = ________

**21.** 57 − 43 = ________

**22.** 39 − 18 = ________

**23.** 76 − 35 = ________

/4

Write these amounts in pounds.

**24.** 312p = £________

**25.** 509p = £________

**26.** 488p = £________

**27.** 620p = £________

/4

Write **symmetrical** or **non-symmetrical** for each of these shapes.

**28.** 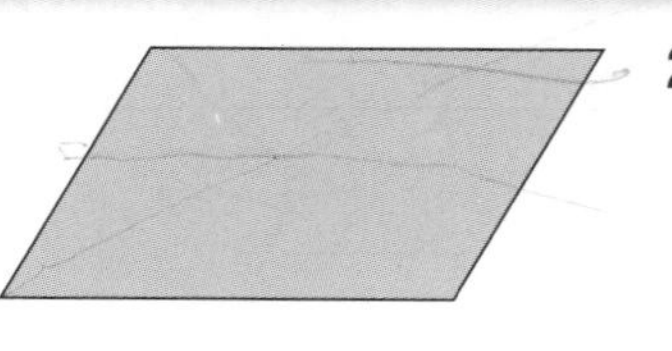 ________

**29.** 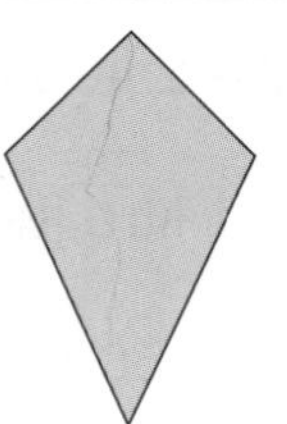 ________

**30.** 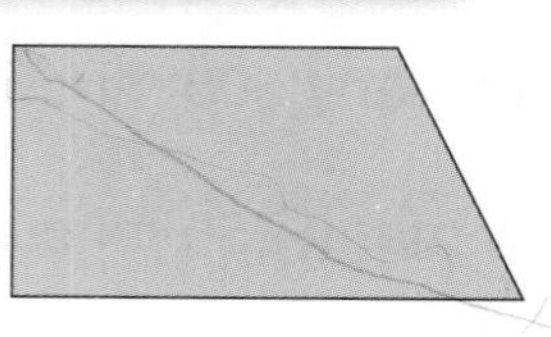 ________

/3

/30

# PAPER 15

Find $\frac{1}{10}$ of each of these.

**1.** 40 → ______ **2.** 80 → ______

**3.** 70 → ______ **4.** 90 → ______

/4

**5.** Tick all the triangular prisms.

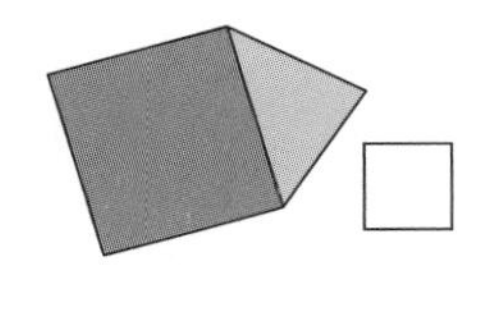

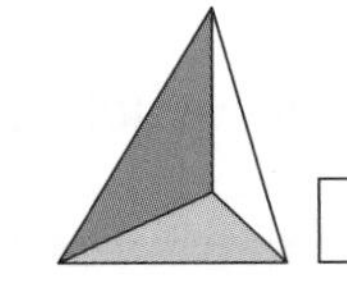

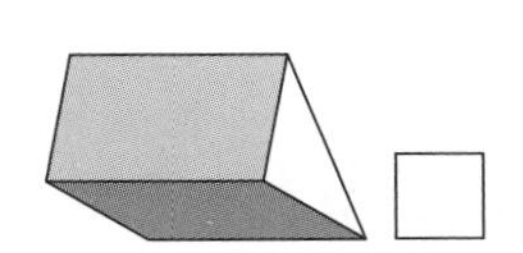

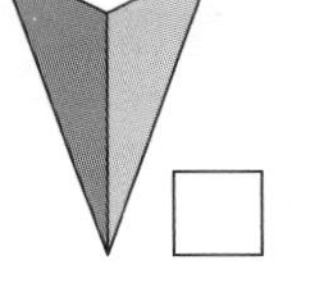

/1

Write the direction you will face for each of these.

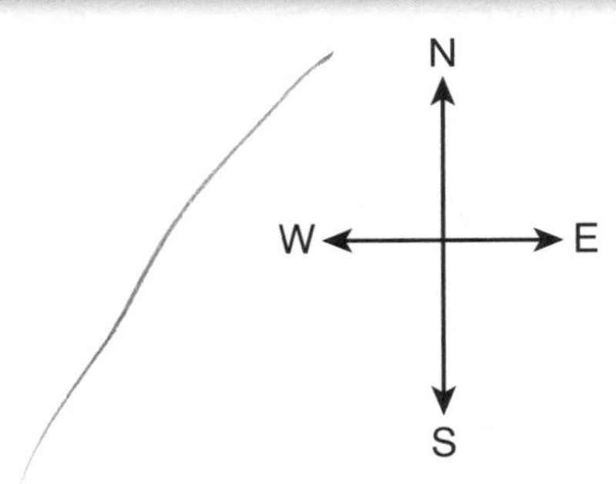


**6.** Face North then turn **clockwise** for 2 right angles. Now facing ________________.

**7.** Face West then turn **clockwise** for 1 right angle. Now facing ________________.

**8.** Face South then turn **anticlockwise** for 1 right angle.
Now facing ________________.

**9.** Face North then turn **anticlockwise** for 1 right angle.
Now facing ________________.

/4

Write < or > for each pair of lengths.

**10.** 180 m ____ 1 km 500 m

**11.** 330 cm ____ 3 m 13 cm

**12.** 2 km ____ 2100 m

**13.** 108 cm ____ 1 m

**14.** 2000 m ____ 1 km 700 m

/5

**15.** **Round** 165 to the nearest 10. Circle the answer.

200 160 170 150

/1

This diagram sorts numbers. Into which boxes will the numbers below be sorted?

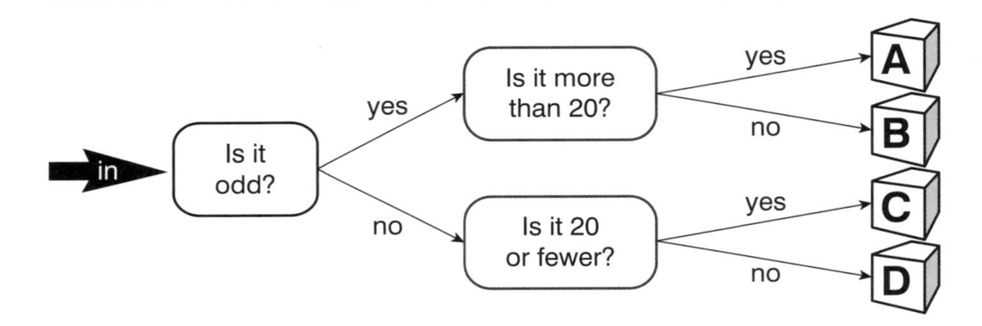


**16.** 18 → ______

**17.** 27 → ______

**18.** 32 → ______

**19.** 15 → ______

**20.** 41 → ______

**21.** 20 → ______

/6

How much liquid is in each container?

**22.**

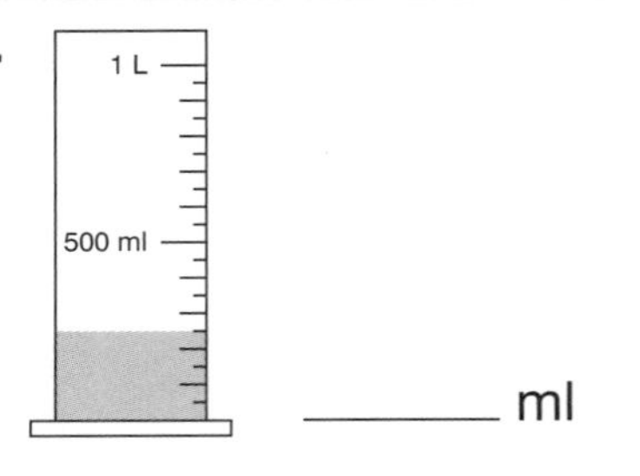


______ ml

**23.**

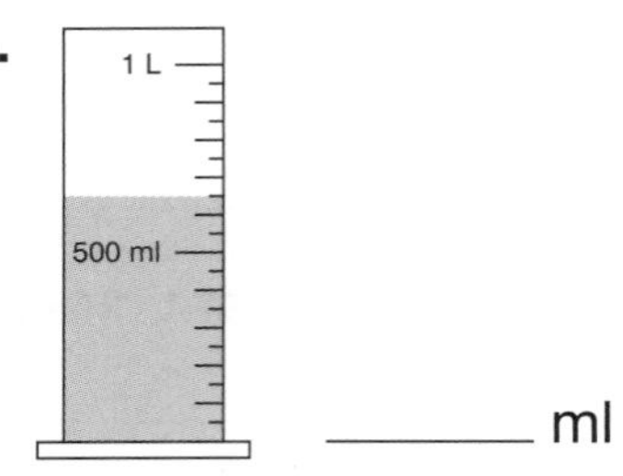


______ ml

/2

**24.** Draw lines to match the pairs that total 100.

| 85 | 74 | 75 | 64 |
|---|---|---|---|
| 25 | 36 | 15 | 26 |

/1

Draw the missing minute hands on these clocks.

**25.**

**6:05**

**26.**

**12:20**

**27.**

**3:10**

**28.**

**9:55**

/4

Write the missing numbers.

**29.** _____ ÷ 5 = 6

**30.** 18 ÷ _____ = 6

/2

/30

## PAPER 16

Write the answers to these.

1. Total 3, 8 and 19 → _____
2. Add together 11, 8 and 5 → _____
3. Total 13, 9 and 6 → _____
4. Add together 18, 14 and 8 → _____

/4

Name these shapes.

5. 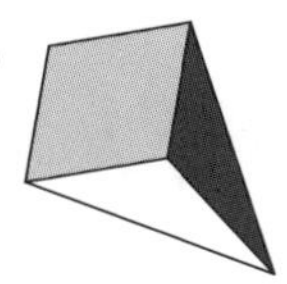 ______________

6. 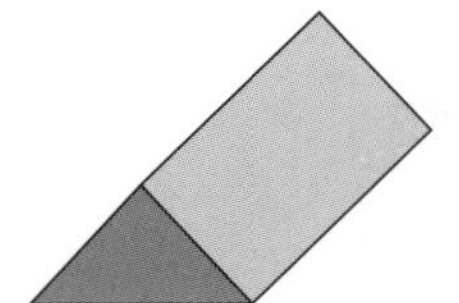 ______________

7. 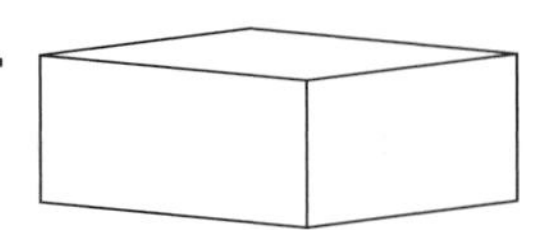 ______________

/3

8. Complete this multiplication grid.

| × | 3 | 10 | 6 |
|---|---|---|---|
| 5 | | | |
| 4 | | | |
| 2 | 6 | | |

/1

Divide each of these numbers by 10.

9. 350 → ______  10. 200 → ______

11. 180 → ______  12. 400 → ______

/4

13. I'm thinking of a number. If I add 12 to it, the answer is 20.

What is my number? ______________

14. I'm thinking of a number. If I subtract 9 from it the answer is 15.

What is my number? ______________

/2

Complete these.

15. 2 weeks = ______ days  16. 120 minutes = ______ hours

17. 3 hours = ______ minutes  18. 21 days = ______ weeks

19. 2 years = ______ months  20. 48 hours = ______ days

/6

Each of these amounts can be made with 4 coins.
Draw the coins.

**21.** £1.80

**22.** £1.26

/2

This graph shows the number of people visiting different shops between 4.00 and 4.30.

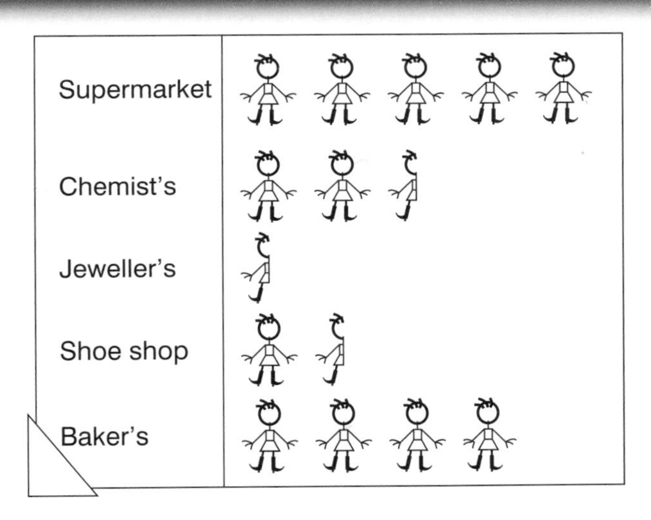


**23.** Which shop was visited by the most number of people? ____________

**24.** Which shop was visited by the fewest number of people? ____________

**25.** How many people visited the Baker's? ____________

**26.** Which shop was visited by 26 people? ____________

/4

Complete these.

**27.** ____ $\times 8 = 16$

**28.** $24 \div 3 =$ ____

**29.** $5 \times$ ____ $= 50$

**30.** ____ $\times 6 = 18$

/4

/30

## PAPER 17

**1.** This machine multiplies numbers by 100.

Complete the table of results.

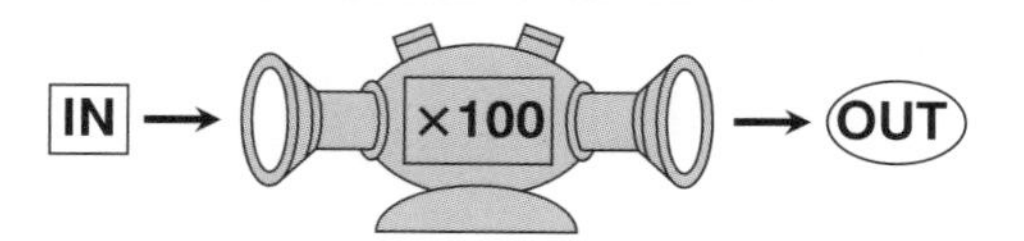


| IN | 7 | | 18 | | 35 |
|---|---|---|---|---|---|
| OUT | | 300 | | 2000 | |

/1

**2.** Tick the shape that has $\frac{3}{4}$ shaded.

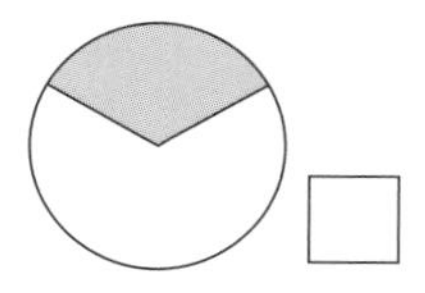 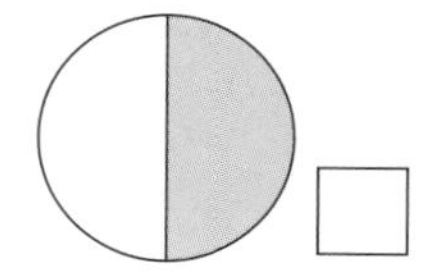 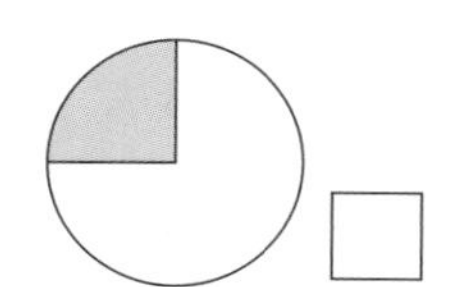 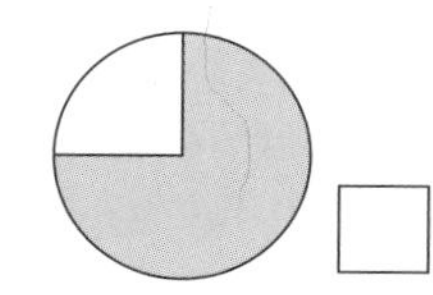

/1

**3.** Write the fractions shaded on these shapes.

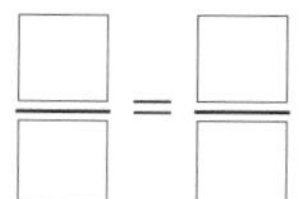

/1

Use the two signs to make these calculations correct.

**4.** 35 + 6 ☐ 50 − 8

**5.** 62 − 4 ☐ 49 + 6

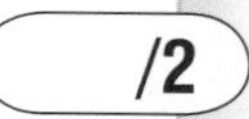


**6.** What is the **area** shaded on this grid? ______ squares

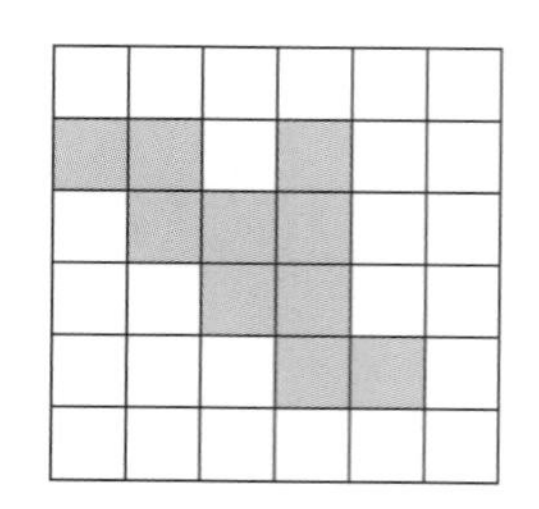

/1

Write the numbers shown on each abacus.

7.
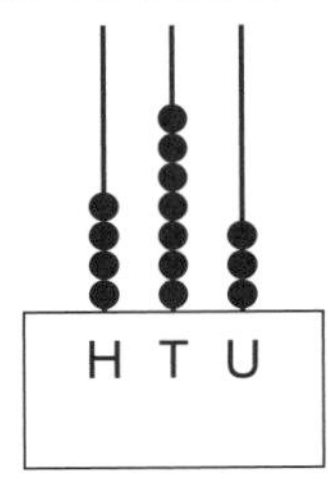


8.
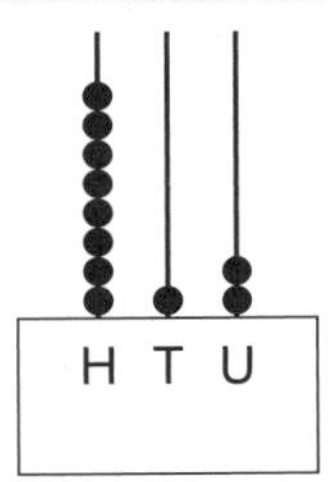


9.
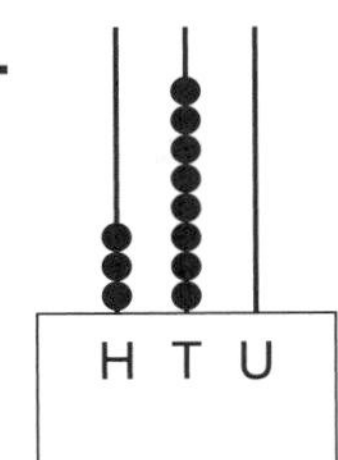


/3

Write the next two numbers in each sequence.

| | | | | | | |
|---|---|---|---|---|---|---|
| **10.** | 940 | 840 | 740 | 640 | ______ | ______ |
| **11.** | 175 | 177 | 179 | 181 | ______ | ______ |
| **12.** | 360 | 350 | 340 | 330 | ______ | ______ |
| **13.** | 172 | 182 | 192 | 202 | ______ | ______ |
| **14.** | 518 | 516 | 514 | 512 | ______ | ______ |

/5

Tick the right angles on these shapes.

15.
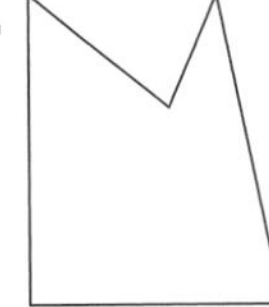

16.
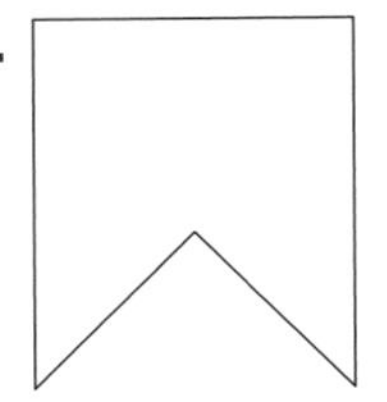

17.
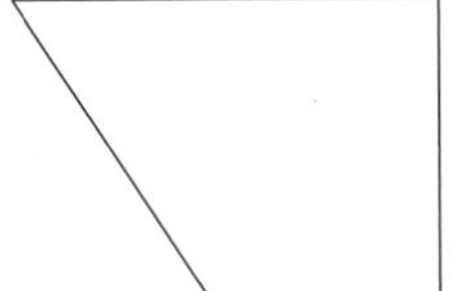

/3

How much liquid is in each of these containers?

18.
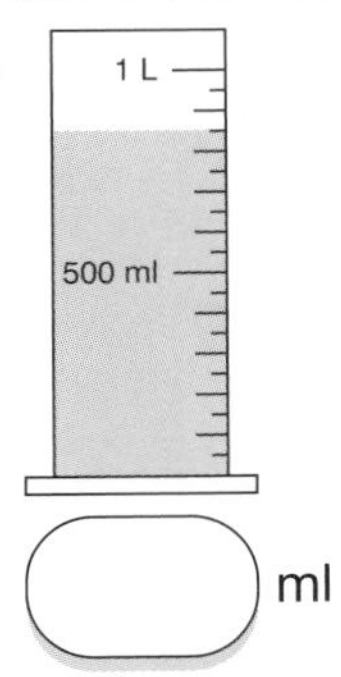


19.
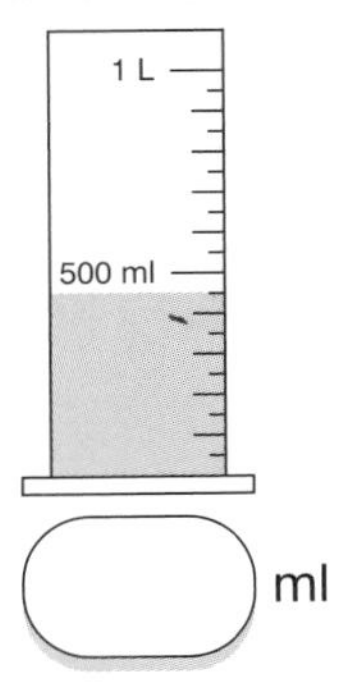


20.
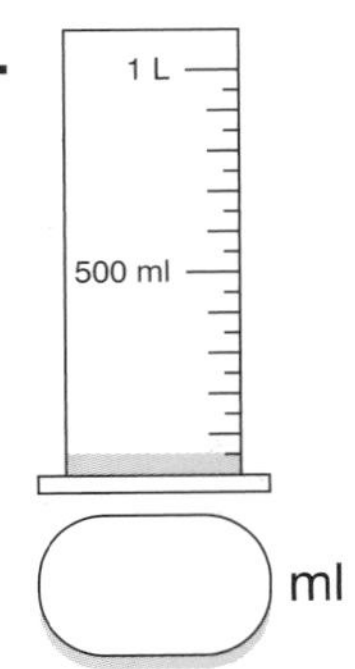


/3

Answer these.

**21.** 43 + 35 = ________

**22.** 45 + 25 = ________

**23.** 32 + 29 = ________

**24.** 28 + 53 = ________

/4

Write the numbers shown on this number line.

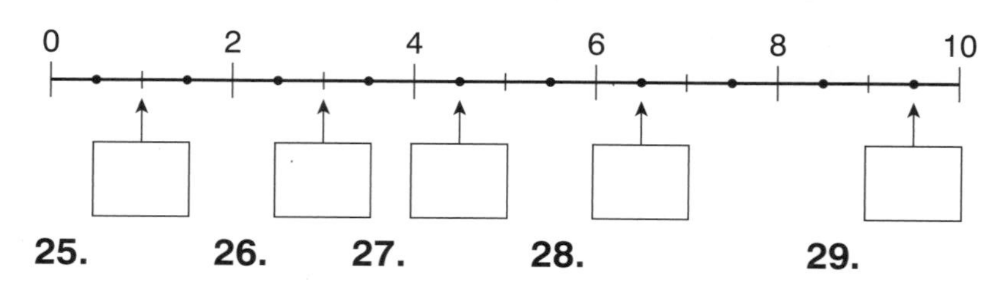


**25.** **26.** **27.** **28.** **29.**

/5

**30.** Tennis balls are put in tubes of 4 balls.
How many tubes can be filled with 25 balls?

________

/1

/30

## PAPER 18

How many grams are the same as these?

**1.** 2 kg = ________ g

**2.** $\frac{1}{2}$ kg = ________ g

**3.** 5 kg = ________ g

**4.** $\frac{1}{4}$ kg = ________ g

**5.** $\frac{3}{4}$ kg = ________ g

/5

Complete these.

6. 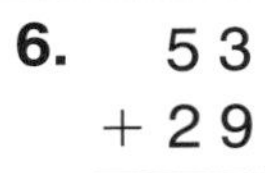

$$\begin{array}{r} 53 \\ +\ 29 \\ \hline \end{array}$$

7.

$$\begin{array}{r} 45 \\ +\ 27 \\ \hline \end{array}$$

8.

$$\begin{array}{r} 37 \\ +\ 48 \\ \hline \end{array}$$

9.

$$\begin{array}{r} 19 \\ +\ 65 \\ \hline \end{array}$$

/4

Name these 3-D shapes.

10. 

____________

11. 

____________

12. 

____________

/3

Start from the centre square for each question.

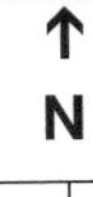


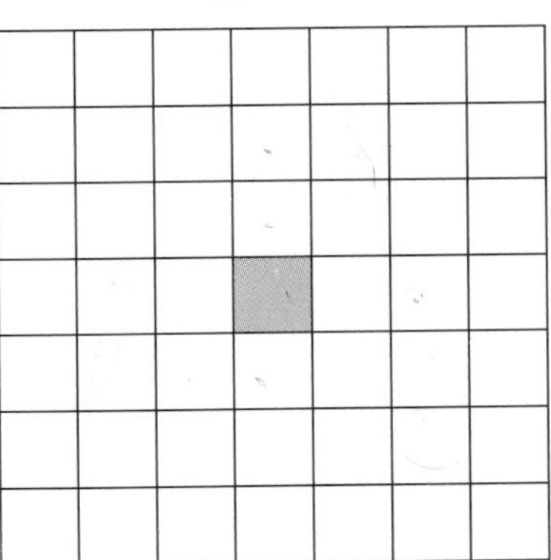

13. Go 2 squares North then 1 square East. Write A in this square.

14. Go 1 square South then 2 squares West. Write B in this square.

15. Go 2 squares East then 2 squares South. Write C in this square.

/3

Read and answer these questions.

**16.** What is the difference between 30 and 46? ______

**17.** Subtract 19 from 45. ______

**18.** What number is 40 less than 58? ______

**19.** What is 80 take away 69? ______

**20.** How much greater is 75 than 66? ______

/5

Draw the lines of symmetry on these shapes.

**21.**

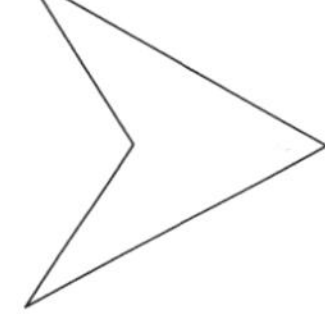

**22.**

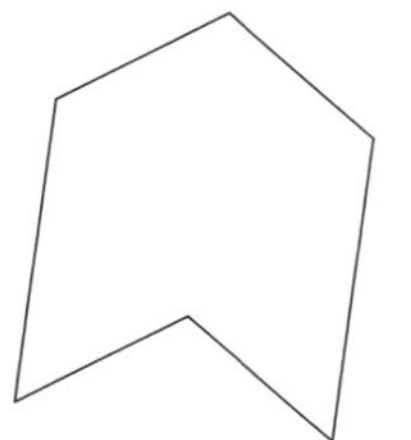

**23.**

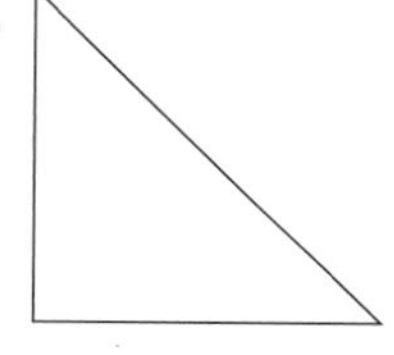

/3

**24.** Circle three numbers that add up to 50.

10 15 35 30 25

/1

Measure the length of each line to the nearest $\frac{1}{2}$ cm.

**25.** Line A = ______ cm

**26.** Line B = ______ cm

**27.** Line C = ______ cm

**28.** Line D = ______ cm

/4

Round these numbers to the nearest ten.

**29.** 345 → ______

**30.** 284 → ______

/2

/30

# PAPER 19

**Write the missing number for each of these.**

**1.** 40 ÷ ____ = 8

**2.** 8 + 15 + ____ = 30

**3.** 6 × ____ = 36

**4.** 38 − ____ = 15

**5.** Circle the number which is one thousand and fourteen.

**a)** 1140 **b)** 1014 **c)** 10014 **d)** 100014

**6.** Two numbers total 85. If the larger number is 61, what is the other number?

________

/2

**Write the next two numbers in these sequences.**

**7.** 45 50 55 60 65 ____ ____

**8.** 120 110 100 90 80 ____ ____

**9.** 30 32 34 36 38 ____ ____

**10.** 59 62 65 68 71 ____ ____

**11.** Tick the right angles in this set.

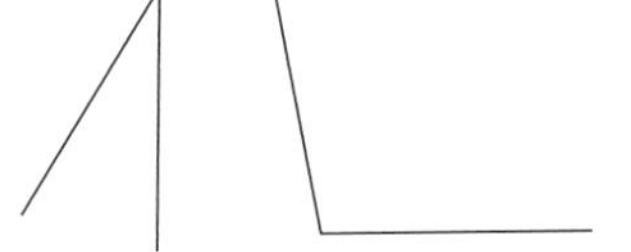

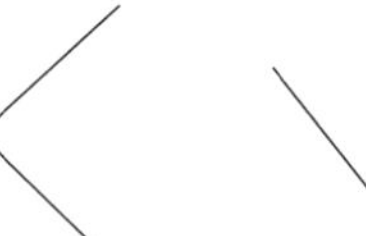

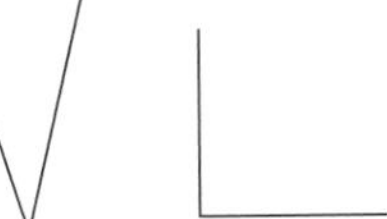

**12.** What is the next odd number after 909? ________

/2

## Complete these.

**13.** $\begin{array}{r} 88 \\ -\,56 \\ \hline \end{array}$ ______

**14.** $\begin{array}{r} 62 \\ -\,49 \\ \hline \end{array}$ ______

**15.** $\begin{array}{r} 77 \\ -\,28 \\ \hline \end{array}$ ______

**16.** $\begin{array}{r} 84 \\ -\,47 \\ \hline \end{array}$ ______

**17.** $\begin{array}{r} 63 \\ -\,45 \\ \hline \end{array}$ ______

**18.** $\begin{array}{r} 52 \\ -\,36 \\ \hline \end{array}$ ______

/6

## Write the missing numbers.

**19.** 486 = 400 + ______ + 6

**20.** 945 = ______ + ______ + 5

**21.** 894 = ______ + 90 + ______

**22.** Round 185 to the nearest 10. Circle the number.

200 180 190 210

/4

**23.** Sort these shapes by writing each letter on the Carroll diagram.

A

B

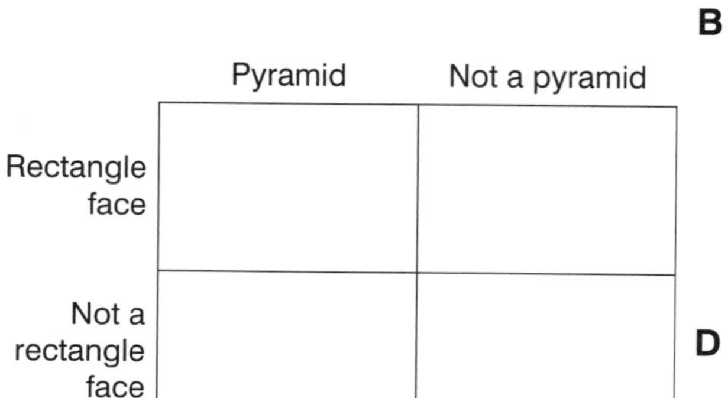


C

D

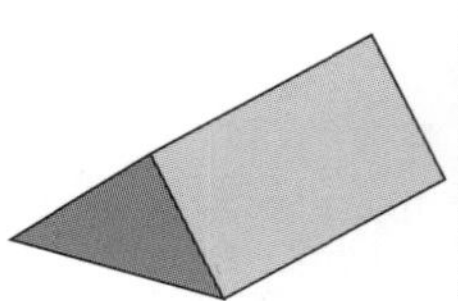

E

F

/1

Complete this table.

| | Length in m and cm | Length in cm |
|---|---|---|
| **24.** | _____ m _____ cm | 110 cm |
| **25.** | 2 m 50 cm | _______ cm |
| **26.** | _____ m _____ cm | 220 cm |
| **27.** | 3 m 40 cm | _______ cm |
| **28.** | _____ m _____ cm | 580 cm |

/5

Answer these.

**29.** $\frac{1}{4}$ of £32 = £__________

**30.** $\frac{1}{3}$ of £18 = £__________

/2

/30

# PAPER 20

Answer these time problems.

1. Jamie gets out of bed at 7.30 and leaves for school 45 minutes later. What time does he leave for school? __________
2. A TV programme starts at 6.15 p.m. and lasts for 55 minutes. What time will it end? __________
3. Mrs Jones drives for 25 minutes to work. If she arrives at 8.30, what time did she set off? __________
4. A train journey lasts for 1 hour 15 minutes. If the train departs at 11.40, at what time will it arrive? __________

/4

## Complete these.

5. ____ ÷ 5 = 7

6. 24 ÷ ____ = 3

7. ____ × 10 = 60

8. 45 ÷ 5 = ____

9. 6 × ____ = 24

10. ____ × 4 = 36

/6

## Draw the reflections of each of these shapes.

11.

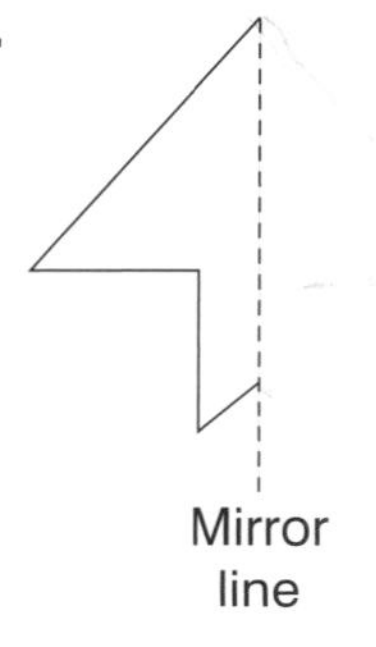


12.

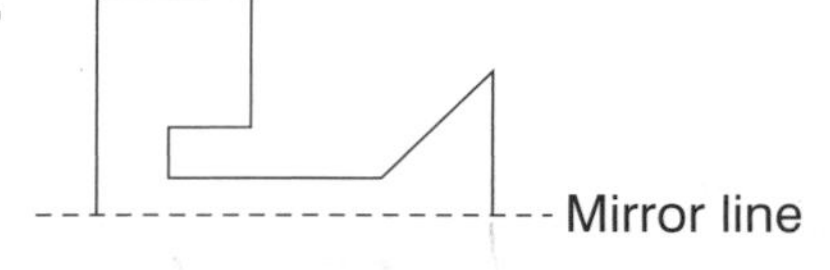


/2

## Answer these.

13. What is 17 added to 34? ____________

14. What is 26 more than 98? ____________

15. What is the total of 38 and 16? ____________

16. What is 34 added to 81? ____________

/4

## **Round** these numbers to the nearest 100.
Draw lines to join them to the correct numbers.

17. 452

18. 634

19. 786

20. 863

400 500 600 700 800 900 1000 1100

/4

Write each group of numbers in order, starting with the smallest.

**21.** 691 528 586 619 ______________________

**22.** 947 940 941 937 ______________________

**23.** 804 840 884 848 ______________________

/3

**24.** Tick the two shapes that have $\frac{1}{3}$ shaded.

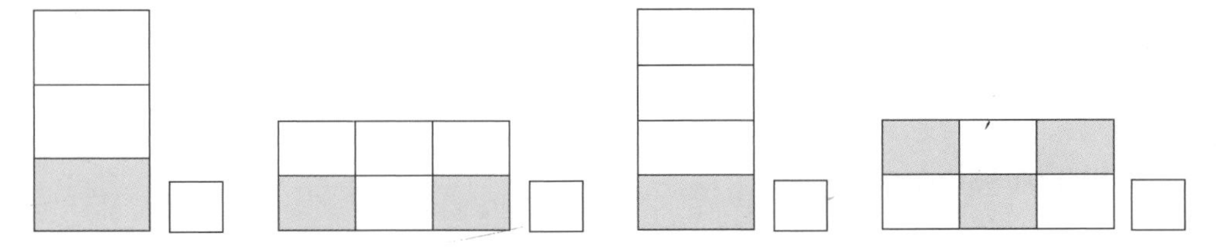

**25.** What shape is this? Circle the answer.

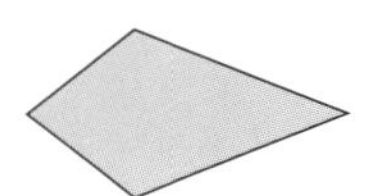

**a)** quadrilateral **b)** pentagon **c)** hexagon **d)** heptagon

Complete these.

**26.** $15 \times 6 =$ ________ **27.** $22 \times 5 =$ ________

**28.** $36 \times 2 =$ ________ **29.** $41 \times 3 =$ ________

**30.** Use a ruler to join dots to draw a right-angled triangle.

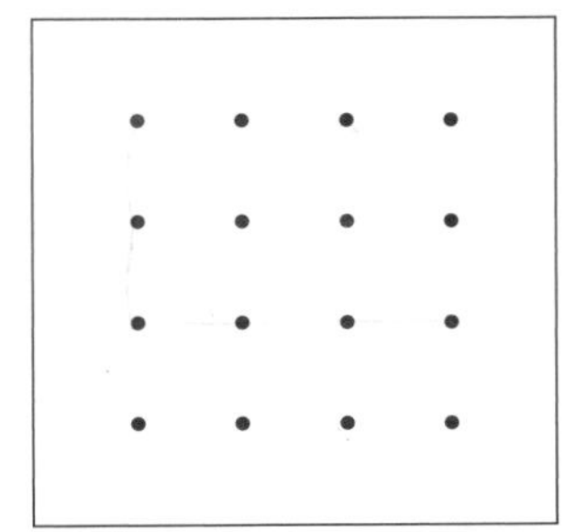

**/30**

# PAPER 21

Complete the table for each multiplying machine.

**1.**

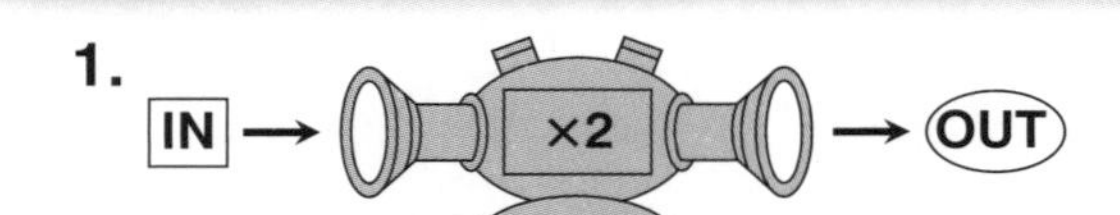


| IN | 2 | | 5 | | 8 |
|---|---|---|---|---|---|
| OUT | | 8 | | 14 | |

**2.**

IN → ×5 → OUT

| IN | 3 | | 6 | | 9 |
|---|---|---|---|---|---|
| OUT | | 25 | | 35 | |

/2

**3.** Circle the missing number.

593 < ______

**a)** 575 **b)** 608 **c)** 499 **d)** 591

/1

Write these times.

**4.**

**5.**

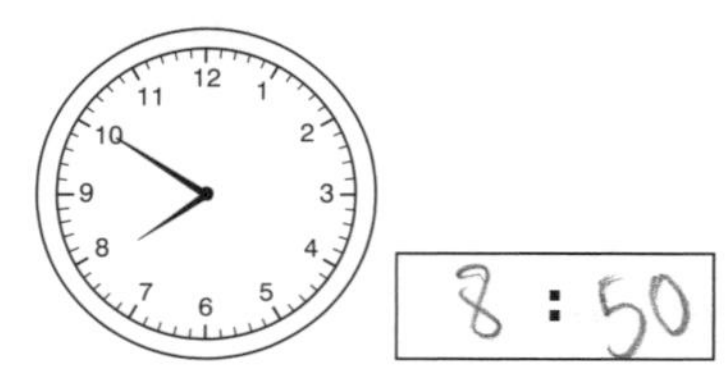


**6.**

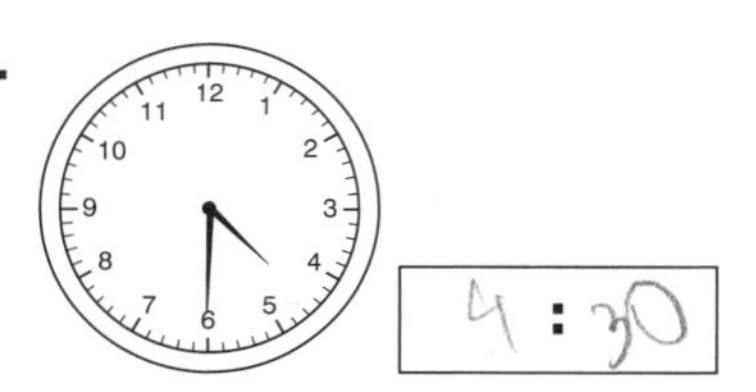


**7.**

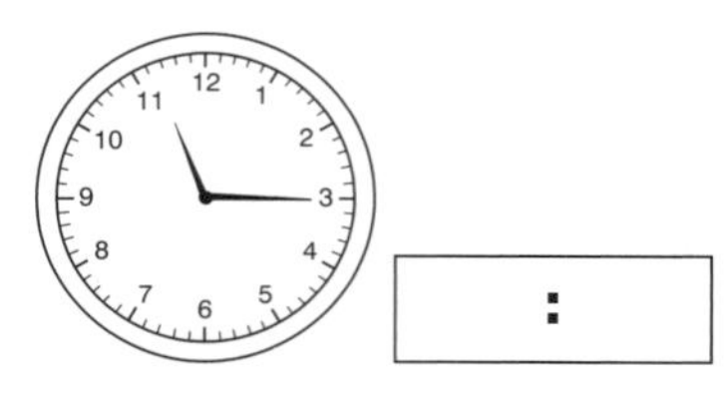

**8.**

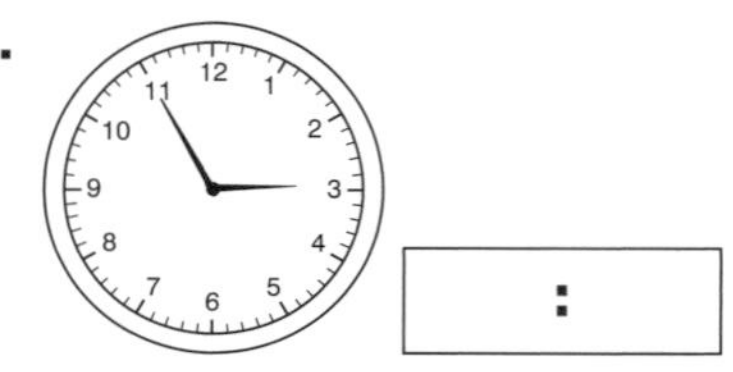

**9.**

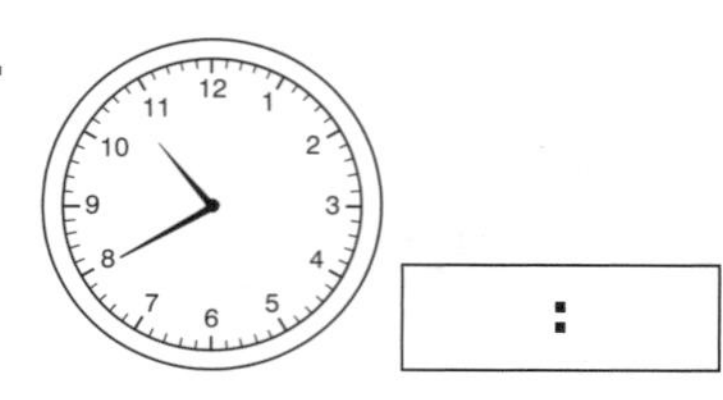

/6

Work out the mystery number for each of these questions.

**10.** When I halve my number and then add 1, the answer is 5.
What is the number? ______

**11.** When I double my number and then subtract 1, the answer is 7.
What is the number? ______

/2

Complete these sums.

**12.**
$$\begin{array}{r} 84 \\ -\ 68 \\ \hline \\ \hline \end{array}$$

**13.**
$$\begin{array}{r} 76 \\ -\ 47 \\ \hline \\ \hline \end{array}$$

**14.**
$$\begin{array}{r} 83 \\ -\ 29 \\ \hline \\ \hline \end{array}$$

**15.**
$$\begin{array}{r} 64 \\ -\ 38 \\ \hline \\ \hline \end{array}$$

/4

Write the missing numbers in these **sequences**.

**16.** 100 ____ ____ 160 180 ____ 220

**17.** 140 130 ____ ____ 100 ____ 80

**18.** 108 ____ 104 102 ____ ____ 96

**19.** 25 ____ ____ ____ 45 50 55

**20.** Write these numbers on the Venn diagram.

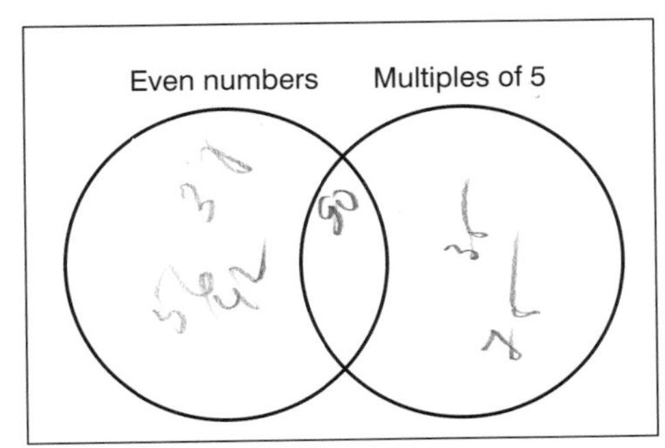

/1

How much liquid is in each container?

21.
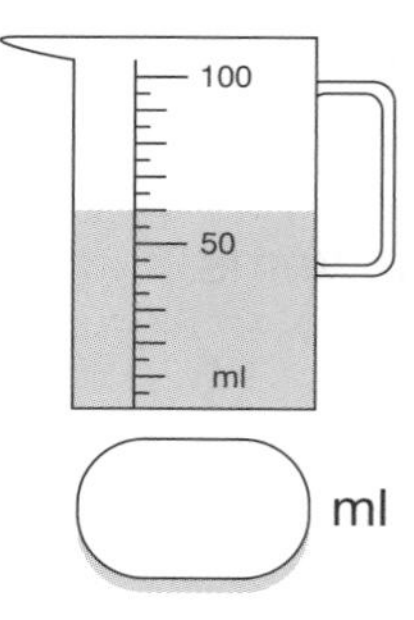


ml

22.
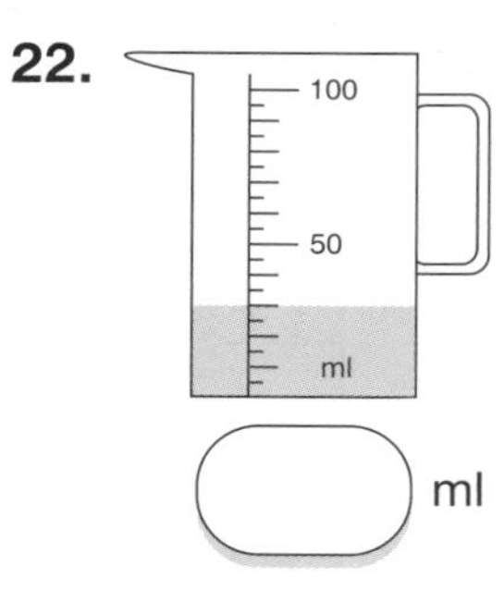


ml

23.
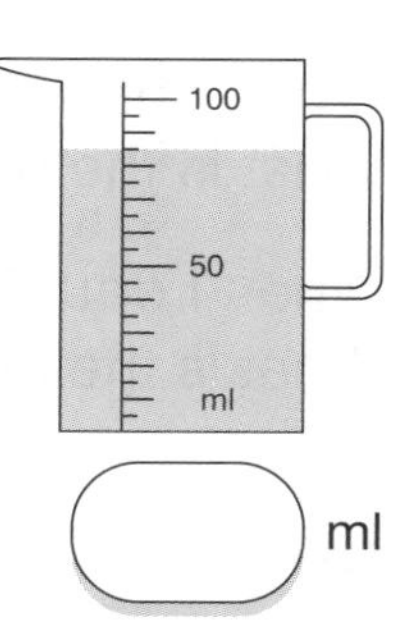


ml

/3

Draw the lines of symmetry on each of these shapes.

24.
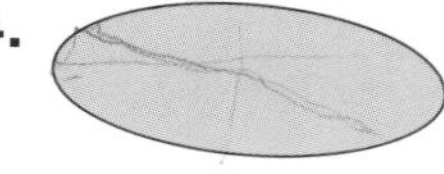

25.
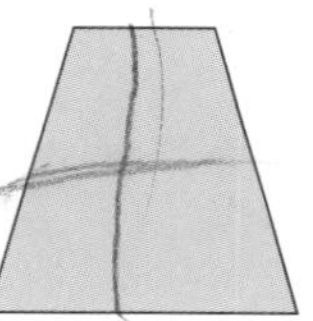

26.
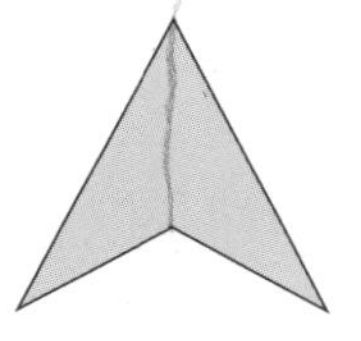

27.
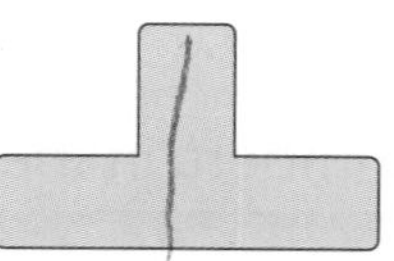

/4

Write the number that is one more than each of these.

28. 109 ________

29. 699 ________

30. 999 ________

/3

/30

# PAPER 22

Complete these.

1. _____ ÷ 2 = 8

2. _____ ÷ 4 = 3

3. _____ ÷ 5 = 3

4. _____ ÷ 3 = 6

/4

5. When one of these numbers is rounded to the nearest 100, the answer is 400. Circle the number.

**a)** 346 **b)** 466 **c)** 454 **d)** 371

/1

Complete the chart below to show the direction you would be facing after each turn.

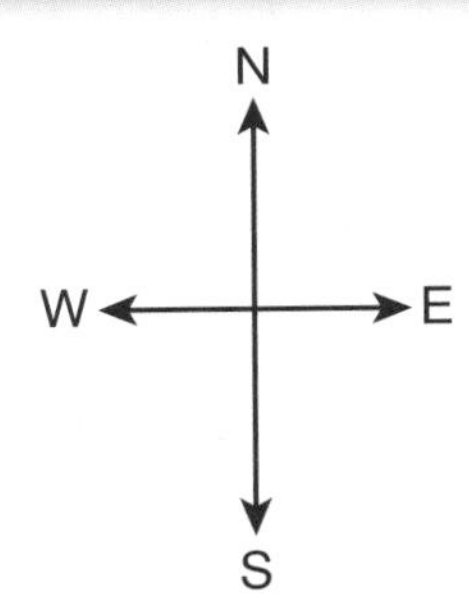


| Start position, facing: | Turn | End position, now facing: |
|---|---|---|
| **6.** North | $\frac{1}{4}$ turn **clockwise** | |
| **7.** South | $\frac{1}{2}$ turn **anticlockwise** | |
| **8.** South | $\frac{1}{4}$ turn **clockwise** | |
| **9.** West | $\frac{1}{4}$ turn **anticlockwise** | |

/4

**Write the missing numbers or words.**

**10.** 984 → nine hundred and ____________________

**11.** ______ → four hundred and seventy-five

**12.** 789 → ____________________

**13.** ______ → eight hundred and nineteen

/4

**How much liquid is in each container?**

**14.**

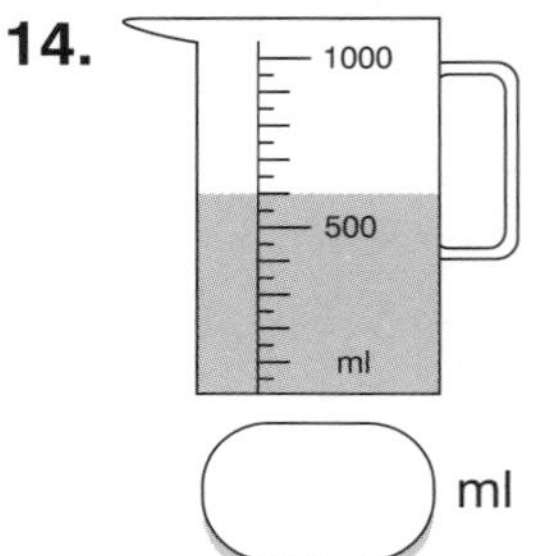


[ ] ml

**15.**

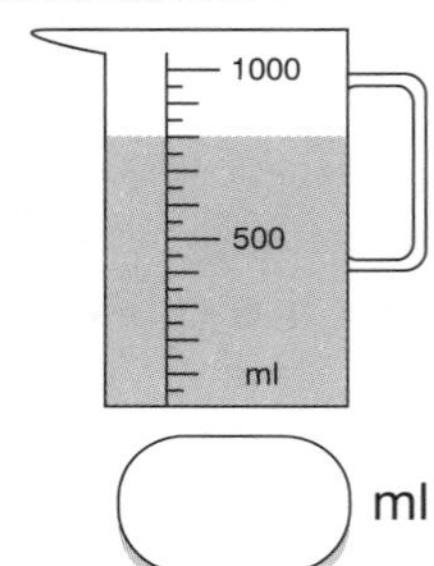


[ ] ml

**16.**

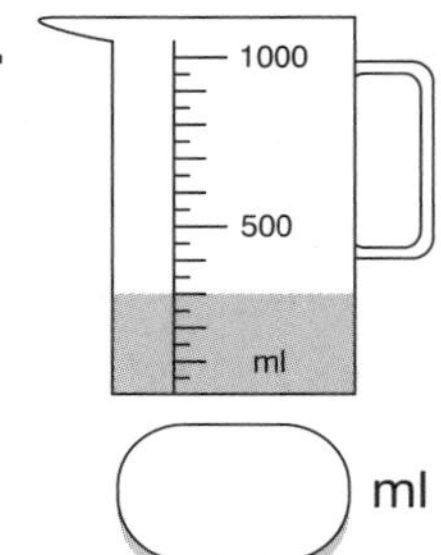


[ ] ml

**17.** Name this shape.

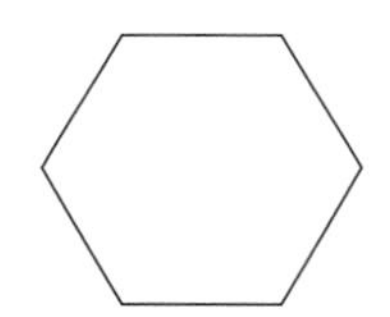

____________________

/4

**Use a ruler to measure the exact length of each line. Write each length to the nearest half centimetre.**

**18.** length: ______cm

**19.** length: ______cm

**20.** length: ______cm

/3

**Write the missing numbers.**

**21.** 45 + ______ = 100

**22.** ______ − 60 = 200

**23.** 250 − ______ = 100

**24.** ______ + 85 = 200

/4

**25.** Tick the pyramids.

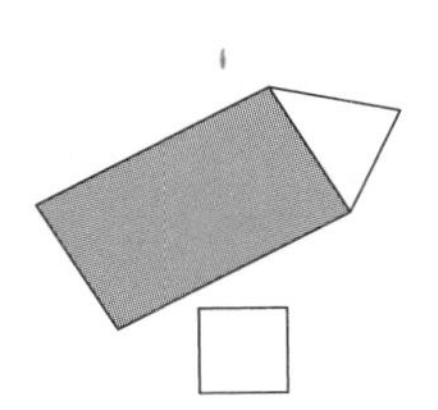
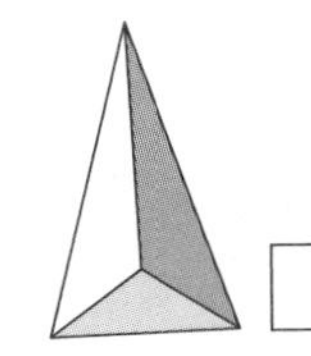
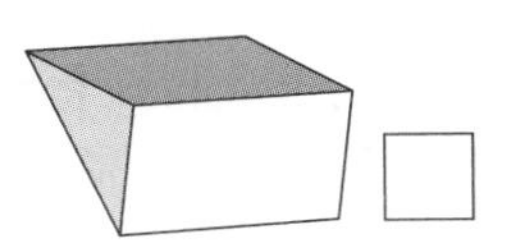
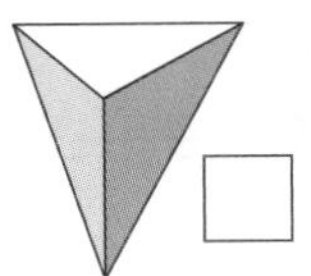
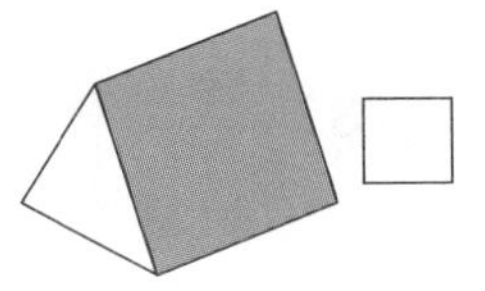

/1

Write the multiplication facts for each number.

**26.**

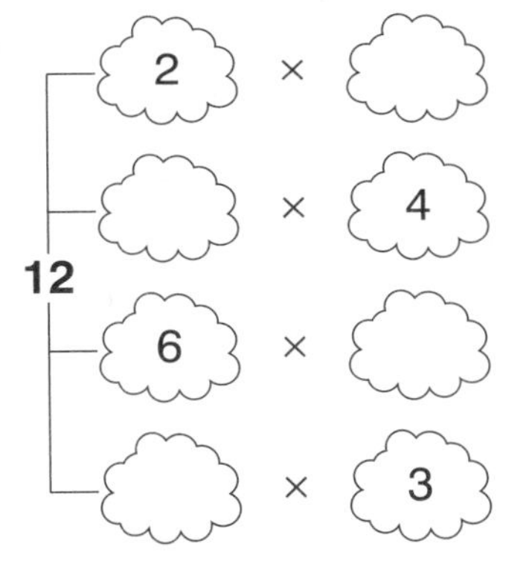


**27.**

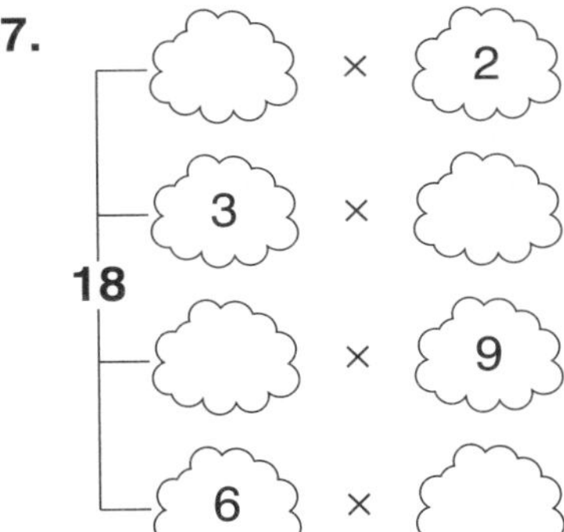


/2

What is $\frac{1}{10}$ of each of these numbers?

**28.** 90 → ____________

**29.** 230 → ____________

**30.** 800 → ____________

/3

/30

# PAPER 23

Write <, > or = for each calculation.

**1.** 9 + 9 ____ 21

**2.** 15 − 8 ____ 7

**3.** 14 + 7 ____ 19

**4.** 23 − 4 ____ 17

**5.** 18 + 11 ____ 33

**6.** 25 − 9 ____ 16

Complete the table for each multiplying machine.

**7.**

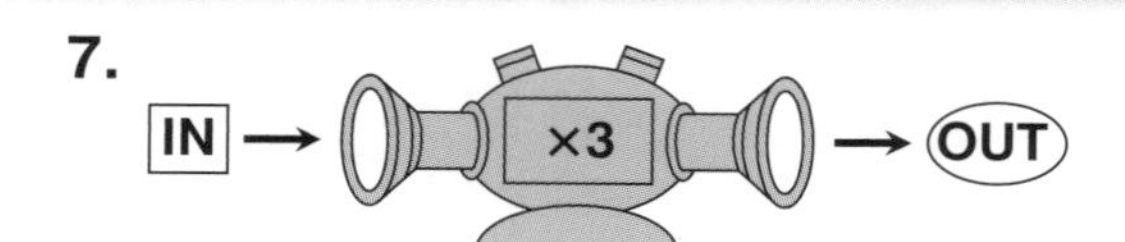


| IN | 2 | | 5 | | 10 |
|---|---|---|---|---|---|
| OUT | | 12 | | 21 | |

**8.**

| IN | 4 | | 7 | | 10 |
|---|---|---|---|---|---|
| OUT | | 20 | | 36 | |

/2

Write the numbers shown on each abacus.

**9.**

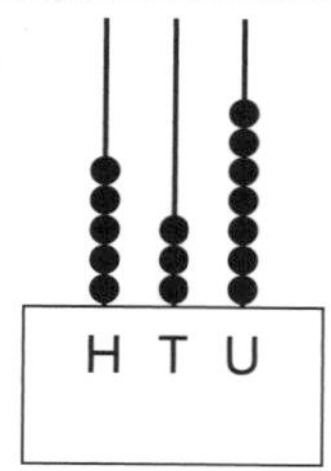


**10.**

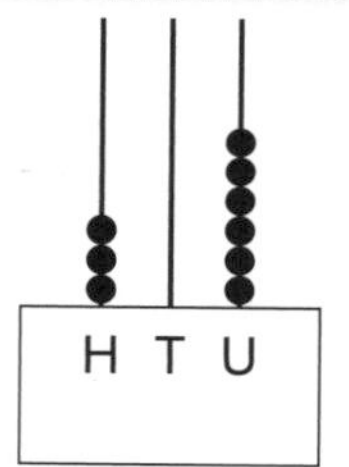


**11.**

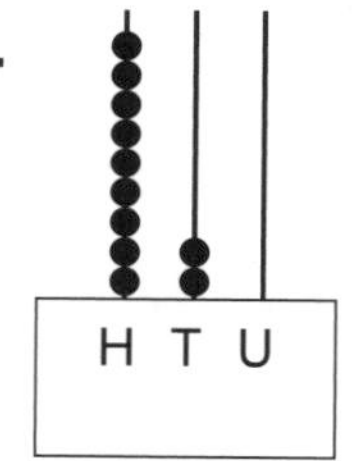


/3

**12.** I'm thinking of a number. If I multiply it by 2, the answer is 20.

What is my number? ____________

Complete these.

**13.** 32 ÷ 4 = ☐

**14.** 25 ÷ 5 = ☐

**15.** 21 ÷ 3 = ☐

**16.** 30 ÷ 6 = ☐

/4

Answer these.

**17.** 5 l = ______ ml

**18.** 600 cm = ______ m

**19.** 7000 m = ______ km

**20.** 3000 g = ______ kg

/4

Look at this 3-D shape.

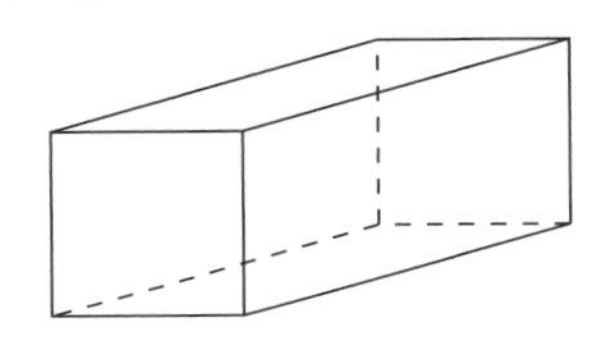

**21.** What is the name of this shape? ______

**22.** How many square **faces** does the shape have? ______

**23.** How many rectangle faces does the shape have? ______

/3

Answer these.

**24.** How many minutes in two hours? ______

**25.** How many days in two weeks? ______

**26.** How many months in two years? ______

**27.** Which day comes after Thursday? ______

**28.** Which month comes after September? ______

**29.** Which of these has a curved face? Circle it.

**a)** cube **b)** cone **c)** prism **d)** pyramid

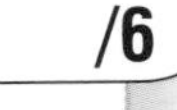

/6

**30.** Circle the fractions that are equivalent to $\frac{1}{2}$.

$\frac{3}{10}$ $\frac{2}{4}$ $\frac{4}{5}$ $\frac{3}{6}$ $\frac{5}{10}$

/1

/30

## PAPER 24

What is the missing number for each of these?

1. 20 ÷ _____ = 4

2. 26 + _____ = 40

3. 3 × _____ = 27

4. 42 − _____ = 33

/4

Complete these.

5.
```
  71
− 38
────

────
```

6.
```
  83
− 57
────

────
```

7.
```
  54
− 28
────

────
```

8.
```
  72
− 49
────

────
```

/4

Start from the centre square each time.

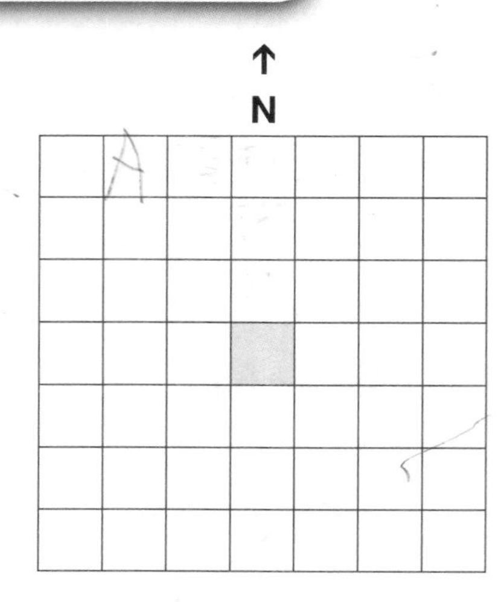


9. Go 3 squares North then 2 squares West. Write A in this square.

10. Go 2 squares South then 3 squares East. Write B in this square.

11. Go 3 squares West then 3 squares South. Write C in this square.

/3

Write each group of numbers in order, starting with the smallest.

**12.** 643 608 650 604 ________________

**13.** 325 319 349 235 ________________

**14.** 536 555 506 560 ________________

**15.** 407 744 474 740 ________________

/4

Complete these.

**16.**

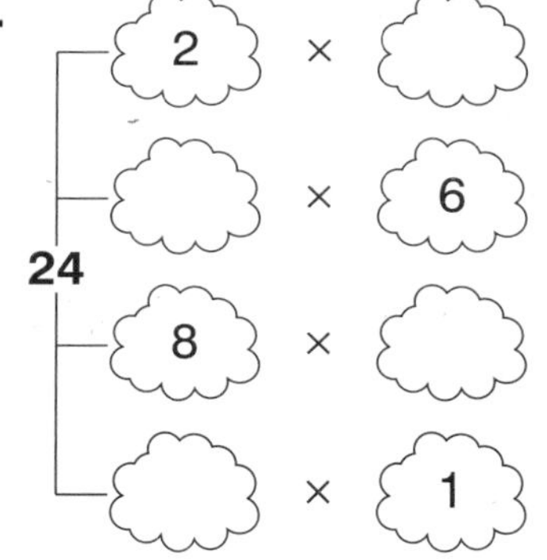


**17.**

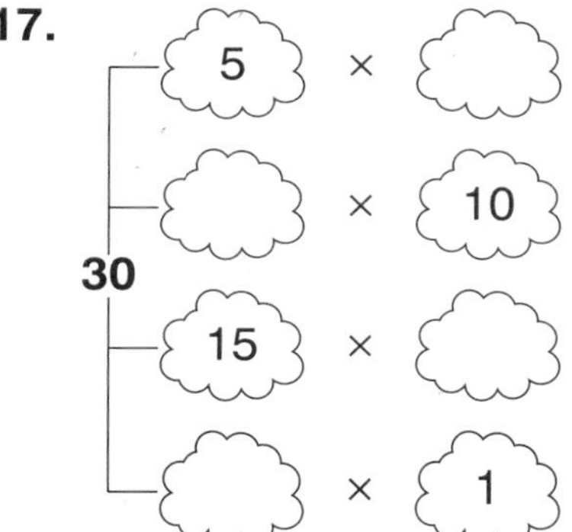


/2

This mapping diagram shows the type of bread a group of children had for packed lunch, and what it was wrapped in.

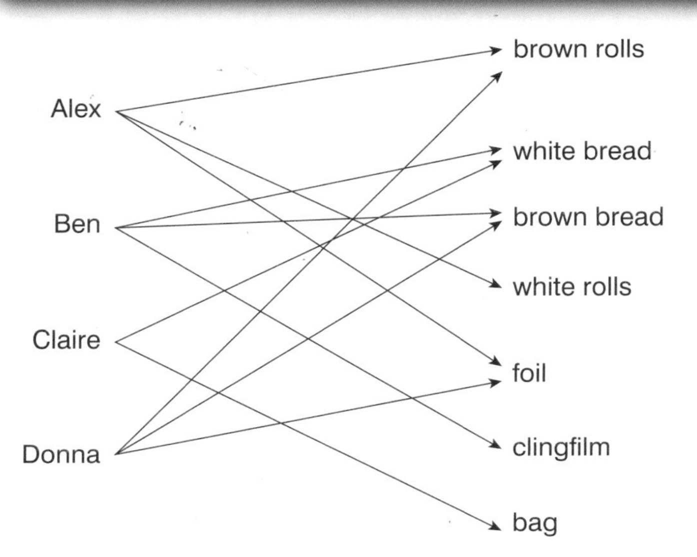

**18.** Who had white bread for their lunch? ____________

**19.** How was Claire's lunch wrapped? ____________

**20.** Who had white rolls for their lunch? ____________

**21.** Did Donna have brown or white rolls? ____________

**22.** Who had brown bread wrapped in clingfilm? ____________

**23.** Who had their lunch wrapped in foil? ____________

**/6**

Name these shapes.

**24.**  ____________

**25.** 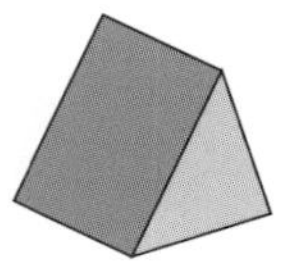 ____________

**26.** 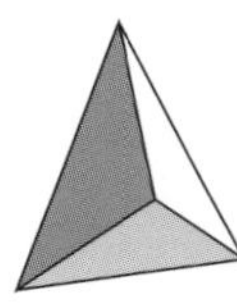 ____________

**27.** 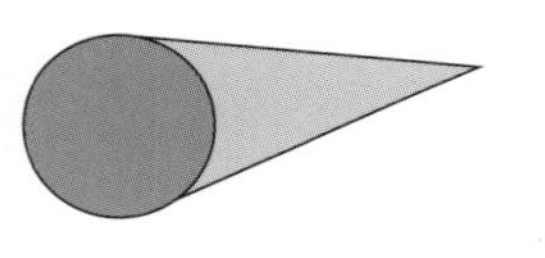 ____________

**/4**

How many minutes are there between these times?

**28.**   ____________

**29.**   ____________

**30.**   ____________

**/3**

**/30**

| | |
|---|---|
| **angle** | the amount by which something turns is an angle. It is measured in degrees (°) |
| **anticlockwise** | turning in this direction |
| **approximate** | a "rough" answer – near to the real answer |
| **area** | the area of a shape is the amount of surface that it covers |
| **axis** (plural: **axes**) | the horizontal and vertical lines on a graph |
| **clockwise** | turning in this direction |
| **denominator** | bottom number of a fraction, the number of parts it is divided into. Example: $\frac{2}{\mathbf{3}}$ |
| **difference** | the difference between two numbers is the amount that one number is greater than the other. The difference between 18 and 21 is 3 |
| **digits** | there are 10 digits : 0 1 2 3 4 5 6 7 8 and 9 that make all the numbers we use |
| **edge** | where two faces of a solid shape meet (edge) |
| **equivalent** | two numbers or measures are equivalent if they are the same or equal |
| **equivalent fractions** | these are equal fractions. Example: $\frac{1}{2} = \frac{2}{4} = \frac{3}{6}$ |
| **estimate** | is like a good guess |
| **faces** | the flat sides of a solid shape (face) |
| **horizontal** | a horizontal line is a straight level line across, in the same direction as the horizon |
| **multiples** | a multiple is a number made by multiplying together two other numbers |
| **non-symmetrical** | when two halves of a shape or pattern are not identical |
| **numerator** | is the top number of a fraction. Example: $\frac{\mathbf{3}}{5}$ |
| **remainder** | if a number cannot be divided exactly by another number then there is a whole number answer with an amount left over, called a remainder |
| **rounding** | rounding a whole number means to change it to the nearest ten, hundred or thousand to give an approximate number. Decimal numbers can be rounded to the nearest whole number, tenth or hundredth |
| **sequence** | a list of numbers which usually have a pattern. They are often numbers written in order |
| **symmetrical** | when two halves of a shape or pattern are identical |
| **vertical** | a line that is straight up or down, at right angles to a horizontal line |
| **vertices** (singular: **vertex**) | These are the corners of 3-D shapes, where edges meet (vertex) |

## Progress grid

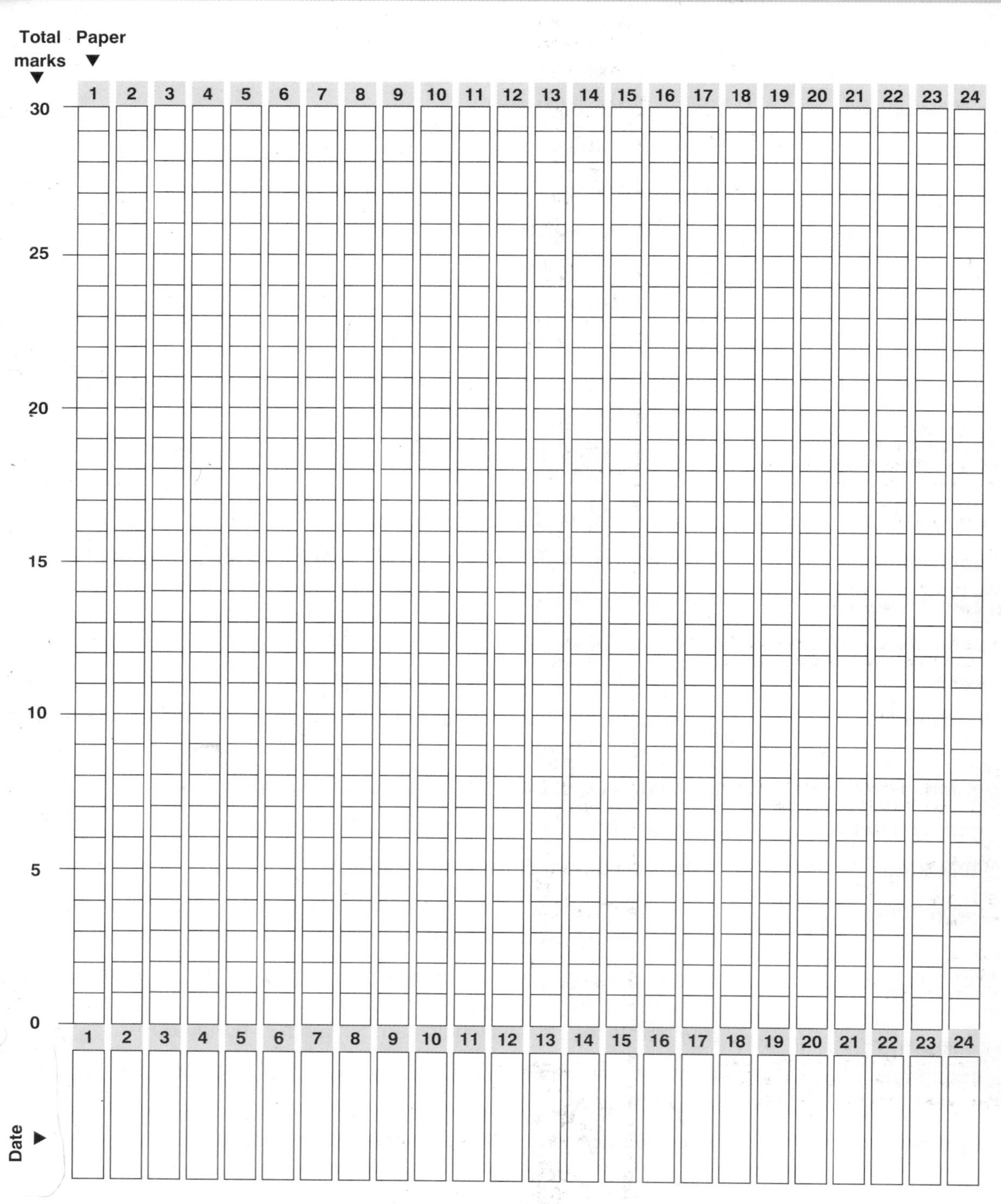


Now colour in your score!